Advances in filament yarn spinning of textiles and polymers

The Textile Institute and Woodhead Publishing

The Textile Institute is a unique organisation in textiles, clothing and footwear. Incorporated in England by a Royal Charter granted in 1925, the Institute has individual and corporate members in over 90 countries. The aim of the Institute is to facilitate learning, recognise achievement, reward excellence and disseminate information within the global textiles, clothing and footwear industries.

Historically, The Textile Institute has published books of interest to its members and the textile industry. To maintain this policy, the Institute has entered into partnership with Woodhead Publishing Limited to ensure that Institute members and the textile industry continue to have access to high calibre titles on textile science and technology.

Most Woodhead titles on textiles are now published in collaboration with The Textile Institute. Through this arrangement, the Institute provides an Editorial Board which advises Woodhead on appropriate titles for future publication and suggests possible editors and authors for these books. Each book published under this arrangement carries the Institute's logo.

Woodhead books published in collaboration with The Textile Institute are offered to Textile Institute members at a substantial discount. These books, together with those published by The Textile Institute that are still in print, are offered on the Elsevier web site at: http://store.elsevier.com/. Textile Institute books still in print are also available directly from the Institute's web site at: www.textileinstitutebooks.com.

A list of Woodhead books on textiles science and technology, most of which have been published in collaboration with The Textile Institute, can be found towards the end of the contents pages.

Woodhead Publishing Series in Textiles: Number 150

Advances in filament yarn spinning of textiles and polymers

Edited by
Dong Zhang

The Textile Institute

ELSEVIER

AMSTERDAM • BOSTON • CAMBRIDGE • HEIDELBERG • LONDON
NEW YORK • OXFORD • PARIS • SAN DIEGO
SAN FRANCISCO • SINGAPORE • SYDNEY • TOKYO
Woodhead Publishing is an imprint of Elsevier

WP
WOODHEAD PUBLISHING

Published by Woodhead Publishing Limited in association with The Textile Institute
Woodhead Publishing is an imprint of Elsevier
80 High Street, Sawston, Cambridge, CB22 3HJ, UK
225 Wyman Street, Waltham, MA 02451, USA
Langford Lane, Kidlington, OX5 1GB, UK

British Library Cataloguing-in-Publication Data
A catalogue record for this book is available from the British Library

Library of Congress Control Number: 2013955406

ISBN 978-0-85709-499-5 (print)
ISBN 978-0-85709-917-4 (online)

For information on all Woodhead Publishing publications
visit our website at http://store.elsevier.com/

Typeset by Newgen Knowledge Works Pvt Ltd, India

Printed and bound in the United Kingdom

Contents

Contributor contact details

(* = main contact)

Editor

D. Zhang
Textile Research Associates
105 Shenandoah Street
Clarksville, TN 37043
USA

E-mail: dzhang@charter.net;
dzhang@textileresearch.org
http://www.t1extileresearch.org/

Chapter 1

V. Kandagor and G. Bhat*
Department of Material Science and Engineering
University of Tennessee
Knoxville, TN 37996, USA

E-mail: gbhat@utk.edu

Chapter 2

E. Devaux
Ecole Nationale Supérieure des Arts et Industries Textiles (ENSAIT)
Laboratoire de Génie et Matériaux Textiles (GEMTEX)
2, allée Louise et Victor Champier
BP 30329
59056 Roubaix Cedex 01, France

and

Centre Européen des Textiles Innovants (CETI)
41, rue des Métissages
CS 40303
59335 Tourcoing, France

E-mail: eric.devaux@ensait.fr

Chapter 3

J. Hagewood
J Hagewood LLC
209 Connors Cir
Cary, NC 27511, USA

E-mail: john@jhagewood.com

Chapter 4

A. Rawal* and S. Mukhopadhyay
Department of Textile Technology
Indian Institute of Technology Delhi
Hauz Khas
New Delhi 110016, India

E-mail: amitrawal77@hotmail.com

Chapter 5

C. J. Kuo* and W. L. Lan
Department of Materials Science and Engineering
National Taiwan University of Science and Technology
43, Sec. 4, Keelung Rd.
Da'an Dist.
Taipei City 106, Taiwan

E-mail: jeffreykuo@mail.ntust.edu.tw

Chapter 6

S. Mukhopadhyay
Department of Textile Technology
Indian Institute of Technology Delhi
Hauz Khas
New Delhi 110016, India

E-mail: samrat@textile.iitd.ac.in

Chapter 7

Q. Q. Ni*, X. D. Jin, H. Xia and F. Liu
Faculty of Textile Science and Technology
Shinshu University
Ueda 386-8567, Japan

E-mail: niqq@shinshu-u.ac.jp

Chapter 8

W. Oxenham
College of Textiles
North Carolina State University
2401 Research Drive
Raleigh, NC 27695-8301, USA

E-mail: william_oxenham@ncsu.edu

Chapter 9

B. Ozipek* and H. Karakas
Istanbul Technical University
Faculty of Textile Technologies and Design
Inonu Cad. No: 65
Gumussuyu/Taksim
Istanbul, 34437, Turkey

E-mail: ozipek@itu.edu.tr

Chapter 10

Y. Imura and M. Jaffe
Department of Biomedical Engineering
New Jersey Institute of Technology
University Heights
Newark, NJ 07102-1982, USA

E-mail: jaffe@njit.edu

R.M.C. Hogan
Omni Tech International
USA

Woodhead Publishing Series in Textiles

1 **Watson's textile design and colour Seventh edition**
Edited by Z. Grosicki
2 **Watson's advanced textile design**
Edited by Z. Grosicki
3 **Weaving Second edition**
P. R. Lord and M. H. Mohamed
4 **Handbook of textile fibres Volume 1: Natural fibres**
J. Gordon Cook
5 **Handbook of textile fibres Volume 2: Man-made fibres**
J. Gordon Cook
6 **Recycling textile and plastic waste**
Edited by A. R. Horrocks
7 **New fibers Second edition**
T. Hongu and G. O. Phillips
8 **Atlas of fibre fracture and damage to textiles Second edition**
J. W. S. Hearle, B. Lomas and W. D. Cooke
9 **Ecotextile '98**
Edited by A. R. Horrocks
10 **Physical testing of textiles**
B. P. Saville
11 **Geometric symmetry in patterns and tilings**
C. E. Horne
12 **Handbook of technical textiles**
Edited by A. R. Horrocks and S. C. Anand
13 **Textiles in automotive engineering**
W. Fung and J. M. Hardcastle
14 **Handbook of textile design**
J. Wilson
15 **High-performance fibres**
Edited by J. W. S. Hearle
16 **Knitting technology Third edition**
D. J. Spencer
17 **Medical textiles**
Edited by S. C. Anand
18 **Regenerated cellulose fibres**
Edited by C. Woodings
19 **Silk, mohair, cashmere and other luxury fibres**
Edited by R. R. Franck
20 **Smart fibres, fabrics and clothing**
Edited by X. M. Tao
21 **Yarn texturing technology**
J. W. S. Hearle, L. Hollick and D. K. Wilson

22 **Encyclopedia of textile finishing**
H-K. Rouette
23 **Coated and laminated textiles**
W. Fung
24 **Fancy yarns**
R. H. Gong and R. M. Wright
25 **Wool: Science and technology**
Edited by W. S. Simpson and G. Crawshaw
26 **Dictionary of textile finishing**
H-K. Rouette
27 **Environmental impact of textiles**
K. Slater
28 **Handbook of yarn production**
P. R. Lord
29 **Textile processing with enzymes**
Edited by A. Cavaco-Paulo and G. Gübitz
30 **The China and Hong Kong denim industry**
Y. Li, L. Yao and K. W. Yeung
31 **The World Trade Organization and international denim trading**
Y. Li, Y. Shen, L. Yao and E. Newton
32 **Chemical finishing of textiles**
W. D. Schindler and P. J. Hauser
33 **Clothing appearance and fit**
J. Fan, W. Yu and L. Hunter
34 **Handbook of fibre rope technology**
H. A. McKenna, J. W. S. Hearle and N. O'Hear
35 **Structure and mechanics of woven fabrics**
J. Hu
36 **Synthetic fibres: nylon, polyester, acrylic, polyolefin**
Edited by J. E. McIntyre
37 **Woollen and worsted woven fabric design**
E. G. Gilligan
38 **Analytical electrochemistry in textiles**
P. Westbroek, G. Priniotakis and P. Kiekens
39 **Bast and other plant fibres**
R. R. Franck
40 **Chemical testing of textiles**
Edited by Q. Fan
41 **Design and manufacture of textile composites**
Edited by A. C. Long
42 **Effect of mechanical and physical properties on fabric hand**
Edited by H. M. Behery
43 **New millennium fibers**
T. Hongu, M. Takigami and G. O. Phillips
44 **Textiles for protection**
Edited by R. A. Scott
45 **Textiles in sport**
Edited by R. Shishoo
46 **Wearable electronics and photonics**
Edited by X. M. Tao
47 **Biodegradable and sustainable fibres**
Edited by R. S. Blackburn
48 **Medical textiles and biomaterials for healthcare**
Edited by S. C. Anand, M. Miraftab, S. Rajendran and J. F. Kennedy

49 **Total colour management in textiles**
Edited by J. Xin
50 **Recycling in textiles**
Edited by Y. Wang
51 **Clothing biosensory engineering**
Y. Li and A. S. W. Wong
52 **Biomechanical engineering of textiles and clothing**
Edited by Y. Li and D. X-Q. Dai
53 **Digital printing of textiles**
Edited by H. Ujiie
54 **Intelligent textiles and clothing**
Edited by H. R. Mattila
55 **Innovation and technology of women's intimate apparel**
W. Yu, J. Fan, S. C. Harlock and S. P. Ng
56 **Thermal and moisture transport in fibrous materials**
Edited by N. Pan and P. Gibson
57 **Geosynthetics in civil engineering**
Edited by R. W. Sarsby
58 **Handbook of nonwovens**
Edited by S. Russell
59 **Cotton: Science and technology**
Edited by S. Gordon and Y-L. Hsieh
60 **Ecotextiles**
Edited by M. Miraftab and A. R. Horrocks
61 **Composite forming technologies**
Edited by A. C. Long
62 **Plasma technology for textiles**
Edited by R. Shishoo
63 **Smart textiles for medicine and healthcare**
Edited by L. Van Langenhove
64 **Sizing in clothing**
Edited by S. Ashdown
65 **Shape memory polymers and textiles**
J. Hu
66 **Environmental aspects of textile dyeing**
Edited by R. Christie
67 **Nanofibers and nanotechnology in textiles**
Edited by P. Brown and K. Stevens
68 **Physical properties of textile fibres Fourth edition**
W. E. Morton and J. W. S. Hearle
69 **Advances in apparel production**
Edited by C. Fairhurst
70 **Advances in fire retardant materials**
Edited by A. R. Horrocks and D. Price
71 **Polyesters and polyamides**
Edited by B. L. Deopura, R. Alagirusamy, M. Joshi and B. S. Gupta
72 **Advances in wool technology**
Edited by N. A. G. Johnson and I. Russell
73 **Military textiles**
Edited by E. Wilusz
74 **3D fibrous assemblies: Properties, applications and modelling of three-dimensional textile structures**
J. Hu

75 **Medical and healthcare textiles**
Edited by S. C. Anand, J. F. Kennedy, M. Miraftab and S. Rajendran
76 **Fabric testing**
Edited by J. Hu
77 **Biologically inspired textiles**
Edited by A. Abbott and M. Ellison
78 **Friction in textile materials**
Edited by B. S. Gupta
79 **Textile advances in the automotive industry**
Edited by R. Shishoo
80 **Structure and mechanics of textile fibre assemblies**
Edited by P. Schwartz
81 **Engineering textiles: Integrating the design and manufacture of textile products**
Edited by Y. E. El-Mogahzy
82 **Polyolefin fibres: Industrial and medical applications**
Edited by S. C. O. Ugbolue
83 **Smart clothes and wearable technology**
Edited by J. McCann and D. Bryson
84 **Identification of textile fibres**
Edited by M. Houck
85 **Advanced textiles for wound care**
Edited by S. Rajendran
86 **Fatigue failure of textile fibres**
Edited by M. Miraftab
87 **Advances in carpet technology**
Edited by K. Goswami
88 **Handbook of textile fibre structure Volume 1 and Volume 2**
Edited by S. J. Eichhorn, J. W. S. Hearle, M. Jaffe and T. Kikutani
89 **Advances in knitting technology**
Edited by K-F. Au
90 **Smart textile coatings and laminates**
Edited by W. C. Smith
91 **Handbook of tensile properties of textile and technical fibres**
Edited by A. R. Bunsell
92 **Interior textiles: Design and developments**
Edited by T. Rowe
93 **Textiles for cold weather apparel**
Edited by J. T. Williams
94 **Modelling and predicting textile behaviour**
Edited by X. Chen
95 **Textiles, polymers and composites for buildings**
Edited by G. Pohl
96 **Engineering apparel fabrics and garments**
J. Fan and L. Hunter
97 **Surface modification of textiles**
Edited by Q. Wei
98 **Sustainable textiles**
Edited by R. S. Blackburn
99 **Advances in yarn spinning technology**
Edited by C. A. Lawrence
100 **Handbook of medical textiles**
Edited by V. T. Bartels
101 **Technical textile yarns**
Edited by R. Alagirusamy and A. Das

102 **Applications of nonwovens in technical textiles**
Edited by R. A. Chapman
103 **Colour measurement: Principles, advances and industrial applications**
Edited by M. L. Gulrajani
104 **Fibrous and composite materials for civil engineering applications**
Edited by R. Fangueiro
105 **New product development in textiles: Innovation and production**
Edited by L.Horne
106 **Improving comfort in clothing**
Edited by G. Song
107 **Advances in textile biotechnology**
Edited by V. A. Nierstrasz and A. Cavaco-Paulo
108 **Textiles for hygiene and infection control**
Edited by B. McCarthy
109 **Nanofunctional textiles**
Edited by Y. Li
110 **Joining textiles: Principles and applications**
Edited by I. Jones and G. Stylios
111 **Soft computing in textile engineering**
Edited by A. Majumdar
112 **Textile design**
Edited by A. Briggs-Goode and K. Townsend
113 **Biotextiles as medical implants**
Edited by M. W. King, B. S. Gupta and R. Guidoin
114 **Textile thermal bioengineering**
Edited by Y. Li
115 **Woven textile structure**
B. K. Behera and P. K. Hari
116 **Handbook of textile and industrial dyeing. Volume 1: Principles, processes and types of dyes**
Edited by M. Clark
117 **Handbook of textile and industrial dyeing. Volume 2: Applications of dyes**
Edited by M. Clark
118 **Handbook of natural fibres. Volume 1: Types, properties and factors affecting breeding and cultivation**
Edited by R. Kozłowski
119 **Handbook of natural fibres. Volume 2: Processing and applications**
Edited by R. Kozłowski
120 **Functional textiles for improved performance, protection and health**
Edited by N. Pan and G. Sun
121 **Computer technology for textiles and apparel**
Edited by J. Hu
122 **Advances in military textiles and personal equipment**
Edited by E. Sparks
123 **Specialist yarn and fabric structures**
Edited by R. H. Gong
124 **Handbook of sustainable textile production**
M. I. Tobler-Rohr
125 **Woven textiles: Principles, developments and applications**
Edited by K. Gandhi
126 **Textiles and fashion: Materials design and technology**
Edited by R. Sinclair
127 **Industrial cutting of textile materials**
I. Viļumsone-Nemes

128 **Colour design: Theories and applications**
Edited by J. Best
129 **False twist textured yarns**
C. Atkinson
130 **Modelling, simulation and control of the dyeing process**
R. Shamey and X. Zhao
131 **Process control in textile manufacturing**
Edited by A. Majumdar, A. Das, R. Alagirusamy and V. K. Kothari
132 **Understanding and improving the durability of textiles**
Edited by P. A. Annis
133 **Smart textiles for protection**
Edited by R. A. Chapman
134 **Functional nanofibers and applications**
Edited by Q. Wei
135 **The global textile and clothing industry: Technological advances and future challenges**
Edited by R. Shishoo
136 **Simulation in textile technology: Theory and applications**
Edited by D. Veit
137 **Pattern cutting for clothing using CAD: How to use Lectra Modaris pattern cutting software**
M. Stott
138 **Advances in the dyeing and finishing of technical textiles**
M. L. Gulrajani
139 **Multidisciplinary know-how for smart textiles developers**
Edited by T. Kirstein
140 **Handbook of fire resistant textiles**
Edited by F. Selcen Kilinc
141 **Handbook of footwear design and manufacture**
Edited by A. Luximon
142 **Textile-led design for the active ageing population**
Edited by J. McCann and D. Bryson
143 **Optimizing decision making in the apparel supply chain using artificial intelligence (AI): From production to retail**
W. K. Wong, Z. X. Guo and S. Y. S. Leung
144 **Mechanisms of flat weaving technology**
V. V. Choogin, P. Bandara and E. V. Chepelyuk
145 **Innovative jacquard textile design using digital technologies**
F. Ng and J. Zhou
146 **Advances in shape memory polymers**
J. Hu
147 **Design of clothing manufacturing processes: A systematic approach to planning, scheduling and control**
J. Gersak
148 **Anthropometry, apparel sizing and design**
D. Gupta and N. Zakaria
149 **Silk: Processing, properties and applications**
K. Murugesh Babu
150 **Advances in filament yarn spinning of textiles and polymers**
D. Zhang
151 **Designing apparel for consumers: The impact of body shape and size**
M.-E. Faust and S. Carrier

Introduction

The use of fibres is one of the most important inventions of human history. From earliest times, people have planted cotton and ramie from which to extract fibres for clothing as a substitute for animal skins, though the quality was often crude and unrefined. The Chinese extracted silk to produce high-quality apparel. The invention of rayon in 1903 triggered a revolution which led to the first synthetic fibre in 1935, so setting in train the rapid development of the synthetic fibre industry. To date, the most commonly used spinning methods include, but are not limited to, melt spinning, dry spinning and wet spinning, with nanofibre technologies emerging more recently.

The history of man-made fibres dates back to the 19th century, and has evolved over the years into a complex field of technology. During the past 50 years, the need for an understanding of the fundamental theory of fibre formation, combined with a market-driven requirement for fibrous materials, has increased interest and attracted attention from scientists, thus enhancing knowledge of these materials.

The inherent qualities of natural fibre-forming materials have limited their useful applications. Silk, cotton and wool have been in use for thousands of years, but their inherent deficiencies (for example, the tendency of cotton to wrinkle, the delicacy required in handling silk, and the shrinking characteristics of wool) limited their application, and the lifetime of such products was therefore short.

In 1903 rayon, the first manufactured fibre, was developed. Advances in the understanding of fibre chemistry for many applications had begun to emerge. By 1950, over 50 different types of man-made polymeric fibres had been produced. This historic advance in fibre technology was preceded by the modification of naturally occurring polymers, enabled by advancing understanding of the background chemistry and physics of polymers.

Since the discovery of natural silk, technology has sought to imitate nature whilst also attempting to produce artificial and synthetic materials with properties similar to those of the natural textile. The exceptional mechanical and handling characteristics of natural silk have remained the reference points in this field. The synthetic fibre invented and patented in 1884 by Bernigaud de Chardonnet, often called 'artificial silk', is considered to be the first significant result in the development of man-made fibres. Development of such materials followed the appearance of plastics at the beginning of the twentieth century and the work done by Hermann Standinger in understanding the macromolecular structure of polymers. Polyamide 6.6 fibre (nylon) developed by DuPont de Nemours in 1938 revolutionised clothing manufacture, as well being used for military applications. The functional characteristics of this fibre closely resemble those of silk. It has exceptional strength, flexibility and lightness and was the first in a long series of synthetic fibres whose properties continued to diversify with progress in macromolecular chemistry.

There are three principal methods of producing polymeric fibres: melt spinning, wet spinning, and dry spinning. Although there are several methods of filament production, melt spinning is the preferred method, and is one of the most economical, mainly due to the simplicity of the process, which does not require solvents. Most commercial synthetic fibres are produced from crystalline polymers by the melt spinning process. For polymers without a stable melt phase, it is necessary to spin the fibre from solution. Dry spinning uses a highly volatile solvent (with a high vapour pressure), so the fibre forms during spinning by solvent evaporation. Wet spinning is used only if the potential polymer solvents are of low volatility. The fibre must be spun in a non-solvent bath to remove the spinning solvent and coagulate the fibre.

As requirements for the performance offered by fibre products increases, the development of reinforced fibre has become a key focus. In the US during the 1970s, DuPont successfully developed an aramid with strength above 17.6cN/dex, leading to the development of high-strength fibres. The main types of high-strength fibres include PPTA fibre, UHMWHP fibre, and carbon fibre.

As the demands made on fibre product performance increased, many special spinning methods were researched and developed. Methods including gel spinning, electrostatic spinning, and integrate composite spinning have been applied in industrial production. Gel spinning has been widely used for the production of fibres with high strength and high mechanical properties. Bi-component fibres are a class of poly-blend fibres, manufactured by special spinning techniques, the aim of which is to overcome the limitations of conventional single-component spinning. The blending of polymers for specialised spinning improves processing and properties for specific end uses.

Electro-spinning is a relatively simple and inexpensive method for producing fibres with diameters in the nanometer range. In recent years, many different kinds of polymer nano-fibres have been produced by this technology. While electrospinning technology is simple, it has a high potential for applications using nano materials. Nano-composite fibres are fabricated by electrospinning technology. These fibres are expected to be multi-functional and are based on the selection of particles and polymers which may have a high potential in engineering applications such as sensing and electroactive actuators. The development of bi-functionality nanocomposite fibres is therefore a challenge for the wider application of nanofibres. However, the production rate for electro-spinning is relatively low when compared to other spinning methods.

Integrated Composite Spinning (ICS) is based on the concept of combining filament yarns with staple fibres to produce a composite yarn. It differs from other yarn formation systems in being designed to integrate a staple fibre component into semi-molten polymer. The concept has aroused considerable interest due to its high production speed when compared to other staple spinning processes. The potential applications for such a system are mainly technical, performance, and medical textiles.

Today, synthetic fibres are ubiquitous. They surround us in clothing, bedding and home furnishings. They are also found in applications such as dust covers for use under sofas and beds, air and liquid filtrations, sound absorbers in car doors, insulation materials in buildings, and as fillings in pillows and cushions. They are primary materials in diapers, cleaning wipes and many window treatments. They are used in dental brushes and floss, and for stitching and bandaging wounds. Many surgical procedures would not be possible without the use of biodegradable synthetic fibre sutures which hydrolyse and dissolve in the body. These fibres are often used as reinforcement materials in plastic parts, automobile tyres, and even in concrete. Much of modern life depends upon our understanding, production and use of synthetic fibres.

Dr Dong Zhang
May, 2013

Part I

General issues

1

Synthetic polymer fibers and their processing requirements

G. BHAT and V. KANDAGOR,
The University of Tennessee, USA

DOI: 10.1533/9780857099174.1.3

Abstract: In order to understand the structure and properties of the desired polymer products, we should have a comprehensive knowledge of the process optimization conditions and materials characteristics. The fundamentals that determine the structure and properties of polymeric fibers include the composition, the molecular structure of the polymer, and morphological features such as crystallinity and orientation. Understanding these factors will be the key to determining correctly the manufacturing process from polymerization to fiber spinning, drawing, and post treatment of manufactured fibers. The inherent qualities of existing natural fiber-forming materials, their limitations and applications, are also discussed.

Key words: polymerization, fiber spinnability, rheology, natural fibers, synthetic fibers, high performance fibers.

1.1 Introduction

The history of man-made fibers dates back to the nineteenth century and has evolved to become one of the most widely-researched fields of technology. During the past 50 years, the need to understand the fundamental theory of fiber formation, combined with the demand for high quality fibers, has attracted the attention of many polymer scientists. For thousands of years, silk, cotton, and wool had been successfully used as textiles, but some drawbacks, for example the tendency of cotton to wrinkle, the delicate handling of silk, and the shrinking characteristics of wool, limited their application. But in 1903, rayon, the first manufactured fiber, was developed. Advances in the understanding of fiber chemistry across a wide spread of applications began to emerge. By 1950, over 50 different types of man-made polymeric fibers had been produced. This historic advance in fiber technology was preceded by the modification of naturally occurring polymers. An advancing and deepening understanding of the background chemistry and physics of polymers has enabled the manipulation of polymers, making possible end products driven by consumer needs.[1]

In order to understand the relationship between processing conditions and the characteristic properties of desired polymer products, a comprehensive knowledge of optimal conditions and material characteristics is necessary. This can be achieved only by understanding the fundamentals of polymers and polymer technology. The fundamentals determining the structure and properties of polymeric fibers include the composition, the molecular structure of the polymer, and morphological features such as crystallinity and orientation. An understanding of these factors is the key to determining the manufacturing process from synthesis of the monomer to polymerization, fiber spinning, and drawing through to the final product.

1.1.1 Types of fiber

Fibers are materials with a very high aspect ratio, or elongated continuous structures, similar to the lengths of thread and varying in diameter from millimeters to nanometers. Most of the commonly used fibers, whether natural or synthetic, tend to be polymeric in nature. Fiber-forming polymers can be categorized as synthetic and natural fibers, and commodity and specialty fibers. Polymers are macromolecular and in their simplest form consist of basic chemical structural units that are identical and linked together by well-defined bonds. The specific type of bond is particularly important in determining the molecular structure and architecture of the polymer. In polymer structure, the number of repeating units in a chain, or the molecular weight of the polymer, is very important in determining its characteristics and applications. The molecular weight of a fiber-forming polymer is not monodispersed, but is rather an average of the molecular weights of the chains that exist in a polymer sample. The molecular weight of the sample is an important property of the polymer, because it determines the tensile strength and influences the physical properties of the material.

Natural and synthetic fibers

Natural fibers originate in geological processes, plants, or animals, and are generally degradable, mainly as a function of time and the environmental conditions to which they are subjected. Synthetic fibers are made from materials that are chemically synthesized and may sometimes imitate natural products. Following World War II, global advances in technology enabled the fabrication of materials equal in strength, appearance, and other characteristics to natural materials. They were also lower in cost. Synthetic fibers can be made from polymers, metals, carbon, silicon carbide, fiberglass, and minerals. This book focuses mainly on polymeric fibers.[1]

The majority of fiber-forming polymers, like common plastics, are based on petrochemical sources. Polymeric fibers can be produced from the following materials: polyamide nylon, polyethylene terephthalate (PET) or polybutylene terephthalate (PBT) polyesters, phenol-formaldehyde (PF), polyvinyl alcohol (PVA), polyvinyl chloride (PVC), and polyolefins (polypropylene (PP) and polyethylene (PE)) among others. Because of the different chemical structures of fiber-forming polymers, their applications vary widely according to the temperature and chemical conditions which they can withstand. For example, polyethylene melts into a viscous liquid at temperatures equal to or less than that of a domestic dryer and therefore its application in a product that will require normal laundering is not possible. However, its fibers can be used in making disposable non-woven products.[2]

Commodity, specialty, and engineered fibers

Most nylons, polyesters and polypropylenes are commodity fibers and have the required properties for apparel and upholstery applications. Specialty polymer fibers are made from plastic materials and have unique characteristics, such as ultra-high strength, electrical conductivity, electro-fluorescence, high thermal stability, higher chemical resistance, and flame retardancy. PE, for instance, a polymer commonly used in the manufacture of disposable shopping bags and refuse bags, is a cheap, low friction coefficient polymer, and is considered a commodity plastic. Medium-density polyethylene (MDPE) is used for underground gas and water pipes; ultra-high molecular weight polyethylene (UHMWPE) is an engineering plastic used extensively for glide rails in industrial equipment and for the low friction sockets in implanted hip joints. It is also used to produce Spectra® or Dyneema®, the fiber having the highest specific strength.

Another industrial application of this material is the fabrication of composite materials consisting of two or more macroscopic phases. Steel-reinforced concrete is one example, and another is the plastic casing used for television sets and cell-phones. This plastic casing is usually of composite materials consisting of a thermoplastic matrix such as acrylonitrile-butadiene-styrene (ABS) to which calcium carbonate chalk, talc, glass fibers, or carbon fibers have been added to enhance strength, bulk, or electro-static dispersion. These additions may be referred to as reinforcing fibers or dispersants, depending on their purpose.

It should be noted here that the dividing line between the various types of plastics is not based on the materials, but rather on their properties and applications.

1.2 Chemistry of fiber-forming polymers

To manufacture useful products from polymers, it is necessary to shape them. This can be achieved by changing the characteristics of a polymer from hard to soft. The two kinds of polymers mainly applied in manufacturing are thermoplastics and thermosets. Thermoplastics, which can easily be melted by subjecting them to the right combination of heat and pressure without necessarily changing the chemical structure, are the most widely used polymers for fiber products. Thermoplastics consist of individual polymer chains that are connected to each other physically rather than chemically. When heated, these chains slide past each other, causing the polymer to become rubbery, eventually causing a flow that enables easier processing. Examples include polyethylene for making milk containers and grocery bags, and polyvinyl chloride for making house wall sidings.

Thermosets consist of interconnected chains with a fixed position relative to each other. These polymers disintegrate into char when subjected to heat because they cannot flow. Examples of thermosets include epoxy resins, Bakelite, and vulcanized rubber, where the individual polymers are chemically interconnected. The chemical crosslinks prevent reorganization of the polymer when subjected to heat, and break only when the thermal energy exceeds the bonding energy between the crosslinks, thus causing disintegration and charring at higher temperatures.

1.2.1 Polymerization

Polymerization is the process of converting a starting chemical, called the monomer, to a long chain molecule, the polymer, which is the basic structure of almost all fibers. The polymerization process will differ according to the chemical composition/structure of the starting monomer. Figure 1.1 below shows the different kinds of polymerization reactions.[3–5] It is obvious that there are many choices, and selection of the polymerization process depends on several factors. Some of these aspects are discussed in the following sections.

1.2.2 Addition polymerization

Addition polymers, also referred to as chain growth polymers, are normally formed by a chain addition reaction process. When an adjacent monomer molecule reacts with an active site of the monomer, chain addition is said to have occurred. Here, the active site is considered the reactive end of the polymer or monomer, which participates in the polymerization process. A schematic representation of a typical chain addition mechanism is shown in Fig. 1.2.

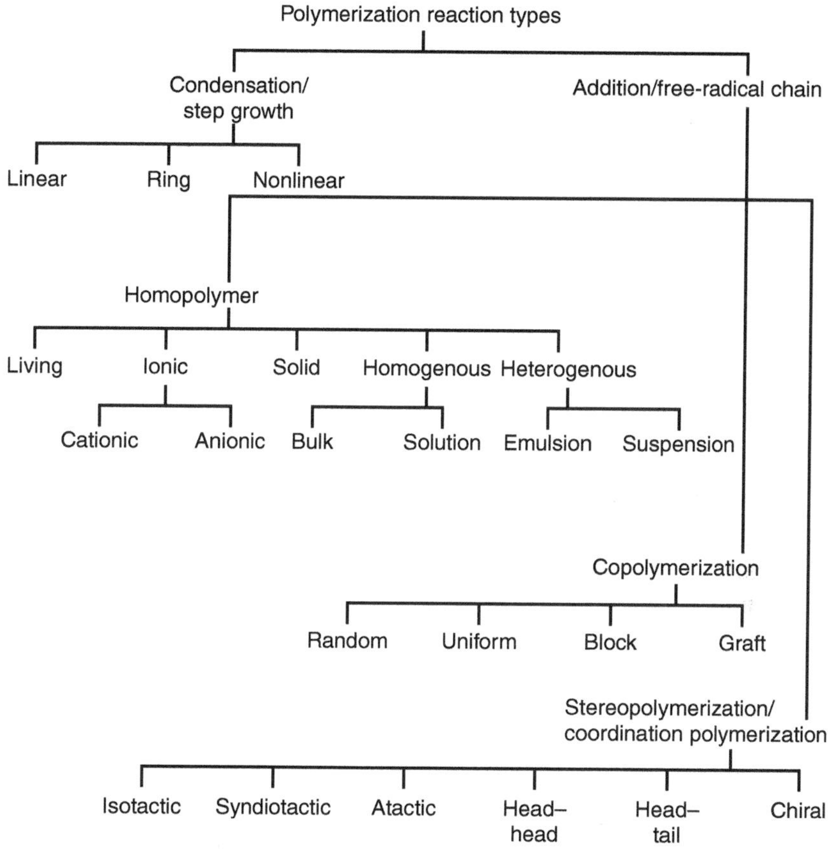

1.1 Polymerization reactions.

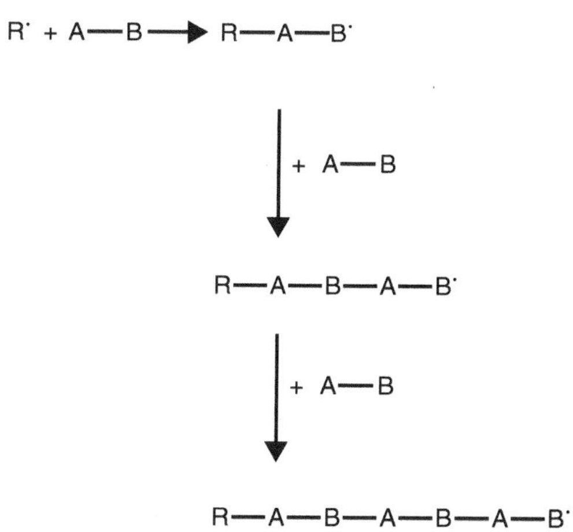

1.2 Schematics of a typical chain addition mechanism.

Commodity polymers, which are typically found in most consumer products, are usually formed by chain growth. These include polyethylene, polystyrene, and polypropylene. Polymerization in chain growth begins with a reaction between a monomer and a reactive species, which results in the formation of an active site. This chain growth polymerization occurs via four mechanisms: anionic, cationic, free radical, and coordination polymerization. These are the most common synthesis mechanisms for the formation of commodity polymers. The details are beyond the scope of this chapter, and can be found elsewhere.[6]

In addition to polymerization, the desired final product may be obtained through three principal steps: (1) initiation: how a polymerization reaction is started; (2) propagation: the polymerization is kept going by the ongoing addition of new monomers to the reactive end; and (3) termination: the means used to stop the reaction. The quality and characteristics of the desired polymer and the choice of the necessary monomer for the reaction process often dictate the choice of the polymerization path. In knowing and understanding the system and the components used in the polymerization process, it is possible to produce a polymer of required molecular weight and molecular weight distribution for specific products.

1.2.3 Condensation polymerization

Polycondensation is the term used to describe polymers formed as a result of reactions involving the condensation of organic materials in which small molecules are split out. In condensation polymerization, molecules of monomer and/or lower molecular weight polymers chemically combine, producing longer chains that are much higher in molecular weight. The polymers usually formed by this mechanism have two functional groups, where functionality is defined as the average number of reacting groups per reacting molecule. The kinetics of polycondensation is usually affected by the formation of a lower molecular component generated during the polycondensation reaction. The molecular component will have a concentration and mass that negatively affects the reaction mechanism. The remedy is to perform the reaction at a higher temperature and to maintain a deep vacuum, which will efficiently and effectively remove the by-products generated during the reaction, thereby favoring the production of higher molecular weight polymer. Polymers typically formed by the polycondensation reaction mechanism include polyesters, nylons, and polyurethanes.[7]

1.2.4 Polymers for specialty and high performance fibers

High performance polymers are slowly replacing traditional materials because of their strength-to-weight ratios and other economic advantages over aluminum or other metals. Polymers used in military applications and the aviation and automobile industries are designed to resist a wider range of thermal, chemical, and physical stress. The orientation of the fibers or polymers for this kind of application is such that it will exhibit superior strength and stiffness to polymers used in conventional applications. The success in developing high performance polymer fibers lies in a deeper understanding of non-crystalline polymer matrix.[8]

Glass fibers have been commercially manufactured since the 1930s and have found relevant application in insulation, fire-resistant fabrics, and reinforcement for fiberglass composites such as bathtub enclosures and boats. Carbon fibers are most widely used in military applications, especially for aircraft and other armored applications, because they can be engineered for strength and stiffness, with varied electrical conductivity, thermal, and chemical properties. These factors are determined by the orientation of the layered carbon planes.[9]

Organic fibers are generally characterized by high modulus and tensile strength. They are used in textiles, aerospace, mechanical, biomedical, geotextile, and electrical applications because of their performance in high thermal environments and stability over a wide range of chemical environments when compared to other conventional fibers. These specialty organic fibers generally possess one or more of the following properties: excellent fire resistance, suitable chemical resistance, higher thermal resistance, wear resistance, increased modulus and strength, and very low density. These polymers typically have an aromatic or cyclic structure, and most of them are produced by polycondensation type reactions.[10,11]

1.2.5 Liquid crystalline/rigid rod polymers

Liquid crystal polymers (LCPs), also known as mesophases, were initially traced in low molar mass compounds over a century ago. LCPs gained recognition in the technological field because of their unique electro-optical properties. Macromolecule LCPs consist of highly ordered oriented polymer chains that introduce crystal-like orientation in the liquid phases. Although the orientation can be very subtle on the local scale, time-averaged fluids are anisotropic and therefore uniquely different from the same attributes that exist in ordinary anisotropic fluids. Figures 1.3 and 1.4 below give a schematic representation of the difference between the melt of a conventional random coil polymer (top) and that of a liquid crystalline polymer.[12]

LCPs have recently become increasingly popular in specialty and high performance applications, and are categorized into two types depending on the principal means of achieving fluidity. Lyotropic LCPs originate from the action of a solvent and are therefore multicomponent polymer solvent

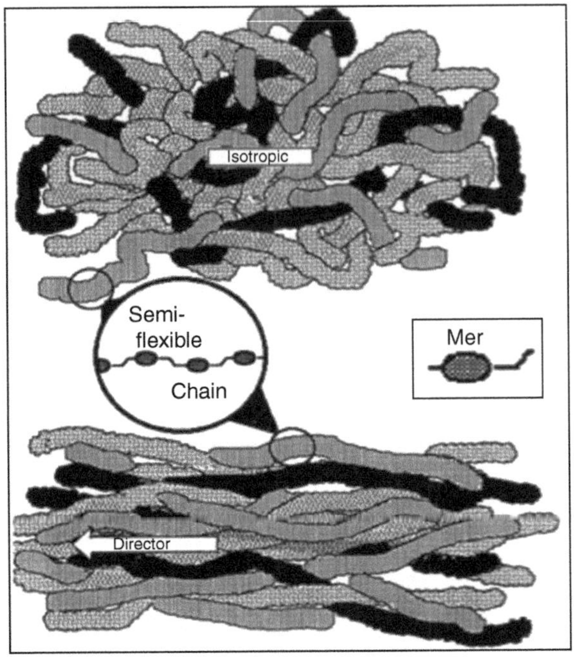

1.3 Schematics difference between the melt of a conventional random coil polymer (top) and that of a liquid crystalline polymer.

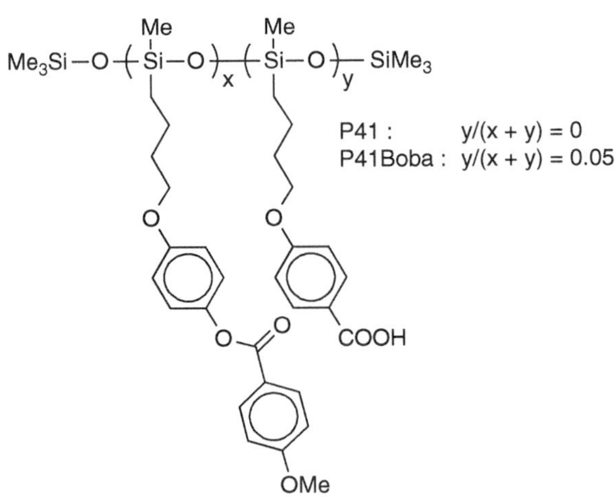

1.4 Structure of a liquid crystalline polymer (*Source*: Copied with permission from the author Polymers 2012, **4**(1), 448–462).

systems. Thermotropic LCPs demonstrate anisotropy in the melt and may be single or multicomponent polymer melts.

The properties of LCPs are uniquely dependent on their final physical forms and the mode of processing. Conventional polymers have been processed by the extrusion of the polymer melt into a stream of evaporating gaseous streams, also known as dry spinning, or by precipitation into liquid medium, which is referred to as wet spinning.

Compared to ultra-high strength conventional polymers, the temperature stability of LCPs is a defining characteristic that makes them effective in many applications. Like other organic polymers, they are characterized by very low densities when compared to inorganic materials such as glass and metals. This means that the tensile strength-to-mass ratios are very favorable for composite applications, and they can therefore be used in aerospace applications.[13]

1.2.6 High strength, high modulus fibers

There is an increasing demand for high temperature fibers, especially in military and aerospace applications where operation in high temperature environments for an extended period of time is critical. Experimental data have shown that incorporating aromatic segments during polymer production significantly increases the thermal stability and melt temperatures of polymers. Most of the synthetic high temperature stable and high strength aromatic fibers are categorized into aromatic polyimides, aromatic polyamides, aromatic polybenzimidazoles, and polybenzimidazolimides. Their structures are shown in Fig. 1.5 below.

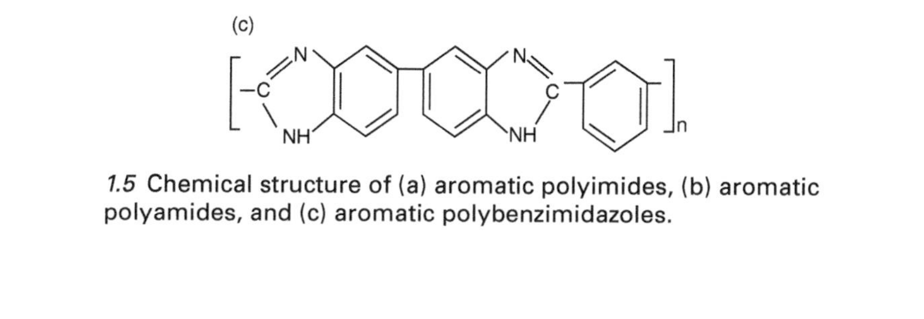

1.5 Chemical structure of (a) aromatic polyimides, (b) aromatic polyamides, and (c) aromatic polybenzimidazoles.

Table 1.1 Fiber properties of the four classes of high temperature synthetic fibers under decomposition in a nitrogen environment

Polymer type	Fiber properties		TGA decomposition temperature (°C)	
	Tenacity (gpd)	Elongation (%)	Air	Nitrogen
Aromatic polyamides	5.5	15	425–450	450–470
Polyimides	7	13	450–500	500–535
Polybenzimidazoles	5	15	300	600
Polybenzimidazolimides	4.5	4	600	600

TGA: thermo-gravimetric analysis

In order to differentiate them from aliphatic polyamides, aromatic polyamides are designated by the generic name 'aramid'. They include three categories based on their composition: aliphatic–aromatic, aromatic–heterocyclic, and wholly aromatic. It has been shown that there is no significant difference in the thermal stability and fiber properties of the four classes of high temperature synthetic fibers. It is only when they decompose in a nitrogen environment that a difference in the thermal properties can be observed in the increasing order of polyamides, polyimides, polybenzimidazoles, and polybenzimidazolimides, as shown in Table 1.1.[14,15]

The term 'high performance,' when used in describing fibers, refers to certain acceptable levels of performance under extraordinary conditions involving temperature, hostile environments, high tensile or compressive exposure, electrical conductivity, or dimensional stability. Since the 1970s, advances in materials engineering technology have led to the development of new synthetic fibers with high strength, high thermal stability, high modulus, chemical and solvent resistance, and low-weight properties. This opens up opportunities in various markets for high performance applications, especially in military, aerospace and several other technological applications. Market penetration for these high performance fibers is increasing exponentially, although volumes are still relatively low. The market is characterized by high prices and complex production processes.

Biopolymers

Polymers from renewable sources offer an alternative solution for maintaining sustainable economic development that will also attract and promote emerging ecologically friendly technologies. Although not all biopolymers are biodegradable, development and innovation in polymers from bio-based materials offers a solution for waste disposal problems. In addition, biological degradability, a drastic reduction in carbon dioxide released into the atmosphere, and increased utilization of agricultural sources for the production of biopolymers will help to reduce concern over the detrimental effect of plastics on the environment. Biodegradable materials are defined as those which degrade through the action of naturally occurring microorganisms, such as bacteria and fungi, over a period of time. Bio-plastics comprise both biodegradable plastics and bio-based plastics. The source of biopolymers is mainly living organisms, and examples include cellulose, starch, chitin, proteins, peptides, and DNA. Most natural biofibers are used as reinforcement by being embedded in another polymeric matrix system that is intended to hold the fibers together, thus providing stability for the required shape in the composite structure. This can transmit the shear forces between mechanically high quality fibers and protect them when subjected to extreme force and vibration.[16,17]

The properties of natural origin polymers can also be chemically or physically modified to allow them to be processed as thermoplastic resins. Polymers from starch and cellulose, for example, may be processed using adjuvants such as water and glycerol. Polylactic acid (PLA), manufactured from completely renewable sources, may be processed into linear aliphatic polyesters and its polymer matrix is compostable. This makes PLA useful in different fiber applications and in packaging, due to its transparency and resistance to water. These properties derive from PLA being made from corn and sugars, which are a 100% renewable source. The formation of PLA is through direct condensation of lactic acid or by cyclic intermediate dimer, a process allowing the removal of water by condensation and by using solvents at high temperature in high vacuum environments.[18]

The properties and characteristics of PLA fibers are very similar to those of many other thermoplastic fibers, especially those of polyester. At room temperature, the polymer is stiff. The glass transition temperature is in the range of 55–65°C. These characteristics and other unique properties, combined with ease of processing, and the fact that PLA can be sourced from renewable products, make it an attractive and acceptable option in several commercial applications. For example, PLA has recently been used for making ski jackets, because of a lower specific gravity and higher resilience than natural fibers, thus making the fabric light with a padded feeling. The potential of PLA extends further than just developing high performance fibers for different applications. It is also a sustainable product, with reduced negative environmental effects than are caused by non-biodegradable polymers. Another polymer currently being developed and marketed by Avantium is the 2,5-Furandi carboxylic acid (FDCA)-based polyester. Although it is in an early stage of its development, this polymer is likely to become another substitute for PET in bottles, fibers, etc.[19–21]

1.3 Polymerization processes

Polymerization is a chemical process in which two or more monomers are combined to form a large repetitive polymer molecule. The polymerization process or path is determined by the required characteristics and application of the end product. If the resulting polymer has more than one molecule in the chain, it is usually referred to as a copolymer. Polymers such as PVC, which consist of repeating long chains of the same monomers, are known as homopolymers. Since the polymerization process can be exothermic in nature, it can result in a hazardous polymerization, which may cause an explosion if the process begins prematurely. Natural polymerization is also known to occur when enzymes polymerase to form nucleic acids, carbohydrates, and proteins.

1.3.1 Criteria for determining the polymerization process

The industrial scale polymerization process is determined by a combination of many factors, both technical and non-technical. A clear understanding of constraints and objectives is needed to choose the right polymerization path. A manufacturer will, for example, determine production rate and the specific chemical and physical properties of the end product. This will determine the molecular weight and molecular weight distribution, copolymer composition and distribution, morphology, particle size distribution, and other factors that will then be used to choose the polymerization process for the desired characteristics. Although polymers are not necessarily sold on the basis of the manufacturing process and structural characteristics, but rather on the end user needs, there remains the challenge of relating the structural features to the properties of the polymer and the polymerization process. The toxicity of the monomers, as well as the flammability of the solvents and other reactants and catalysts necessary for the initiation of polymerization, will also determine the choice of the polymerization process. A clear understanding and choice of the polymerization path will offer the manufacturing industry considerable benefits both in direct and indirect costs. It should be noted that while most initial processes are carried out in a laboratory setting before being implemented commercially, the environment in which industrial scale polymerization is carried out is such that there will naturally be a considerable degree of difference, which will translate into the final product variability.[22]

1.3.2 Bulk polymerization

Bulk polymerization, which may also be referred to as mass polymerization, involves a monomer and an initiator as the main components, without a

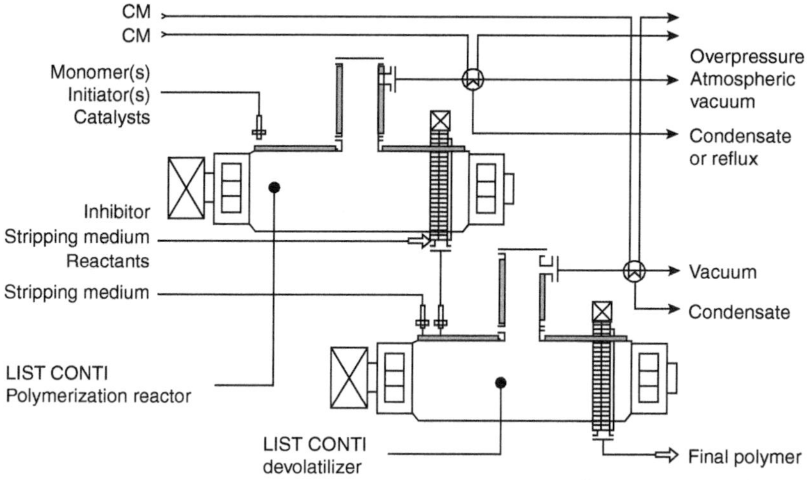

1.6 Kneader reactor concept for bulk polymerization (*Source*: Adapted from Bartke and Diener).[71]

solvent. The polymerization conforms to the fixed shape of the reaction vessel when solidification occurs. The real challenge concerns the dissipation of the heat produced by polymerization. Bulk polymerization is advantageous in the production of cast-molded products such as polystyrene scintillators. Research has shown that to improve the polymerization rate, the monomer should be distilled prior to the reaction and, where this is impracticable, the product should be rinsed with an appropriate solution. Figure 1.6 below shows the schematics of the Kneader reactor concept, which is ideally suited for bulk polymerization.[23,24]

1.3.3 Emulsion polymerization

Emulsion polymerization systems consist of water, up to 3% of surfactants and a free-radical generator soluble in water, rather than being monomer-soluble. The rate of emulsion polymerization is faster than that of the bulk of polymerization, assuming the same monomer is used under the same thermal conditions. The monomer is gradually introduced into the system from the start of the process. The average molecular weight of products obtained by emulsion polymerization is greater and the particle size is usually in the order of a 0.05–0.15 μm. The end product is in the form of emulsion in water, not a filterable suspension. In bulk polymerization by this process, recovery may be done through freezing, heating, or adding salt or acid, but the coagulants, surfactants and the initiators will normally remain as impurities in the final product.[25]

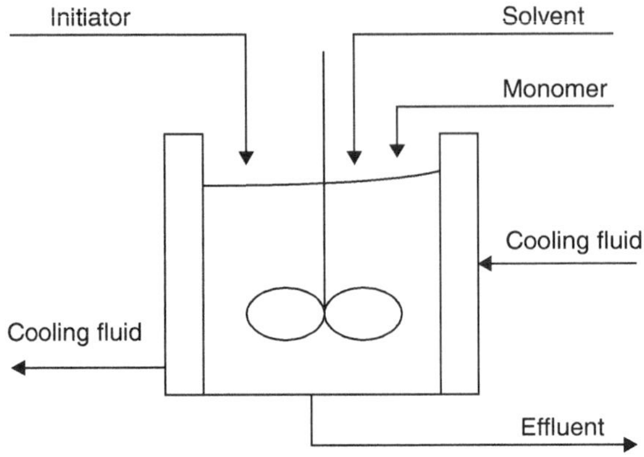

1.7 General schematics of solution polymerization (*Source*: Adapted with permission from Chemical Engineering Journal Volume **143**, Issues 1–3).

1.3.4 Solution polymerization

In this process, the polymerization reaction occurs in a solvent medium that is carefully chosen according to the chemical compatibility of the reacting monomers. Several disadvantages associated with bulk polymerization are resolved by this method. Solution polymerization makes homogeneous polymerization possible, particularly where both the polymer and the monomer are soluble in the solvent. It is also an excellent way of dissipating the reaction heat, reducing shrinkage, and minimizing side reactions in systems that are highly viscous. In ionic polymerization, the solvent medium can positively influence the stereoregularity of the final product. The solvent to be used in solution polymerization should therefore be carefully identified so that its boiling point corresponds to that of the monomer and to the temperatures at which the initiator will decompose. In the industrial production of protective coatings, it is often desirable for the solvent to remain in the paint, and therefore solution polymerization is the process of choice as a strong-polymer interaction is advantageous. Figure 1.7 shows the general schematics of solution polymerization.

1.3.5 Aqueous suspension and irradiation polymerization

In this polymerization path, the monomer droplets are suspended in an aqueous medium and are constantly stirred. Because the monomer droplets are in a water medium, the heat is easily dissipated, making this process comparable to bulk polymerization *in situ*.

Irradiation polymerization is a complex process in which a photo initiator (PI) (usually UV light or electron beam) is used to start the polymerization process. The PI absorbs the incident light in order to generate the ions or reactive radicals which result in polymerization.

1.4 Influence of chemical structure on the physical properties of chemicals and fibers

The general properties of either man-made or naturally occurring polymeric materials are substantially different from those synthesized from metallic or ceramic materials. Polymers have a large molecular weight, which clearly distinguishes them from other materials in which repeated molecules are connected in a different manner from those of the polymer molecule.[26]

Individual polymer characteristics are usually dependent on and defined by the molecular chain architecture, which is dependent on the rotational freedom of atoms around a single bond. This unique aspect of a polymer can be manipulated to obtain the characteristics required for a specific application. Bulky atoms, for example, will greatly reduce the mobility between chain elements because the degree of flexibility of the polymer is reduced – a parameter associated with the transition temperature.[27–29]

It is a fundamental property of polymeric materials that the chemical bonds must be broken and restructured to change the configuration of a particular polymer. In fiber-forming polymers, this is the means by which the desired fiber or fabric qualities, e.g., tensile strength, and stiffness, are tuned. Fibers from polymers are made by spinning or drawing, and certain molecular weight conditions have to be met to achieve predetermined fiber properties. The higher the interchange cohesive forces, for example, the lower the molecular weight at which satisfactory fiber properties are achieved, as these are determined by the molecular weight distribution of a system.[30–34]

Intermediates present during the polymerization process affect the formation of linear polymers. The chemical structure of the intermediates may cause reactions which, over time, will lead to the loss of the intended reactive groups that are important in diminishing chain growth during the polymerization process. These side reactions will also cause chain branching, which may result in cross-linking and the formation of gelled polymers that cannot be spun.

Polymers can exist in the crystalline and amorphous regions, and the crystallinity of polymeric fibers is one of the most important aspects of polymer science. Crystallinity is a measure of the percentage of crystalline regions in the polymer with respect to amorphous region. Figure 1.8 is a schematic representation of polymer chains in the crystalline and amorphous regions.

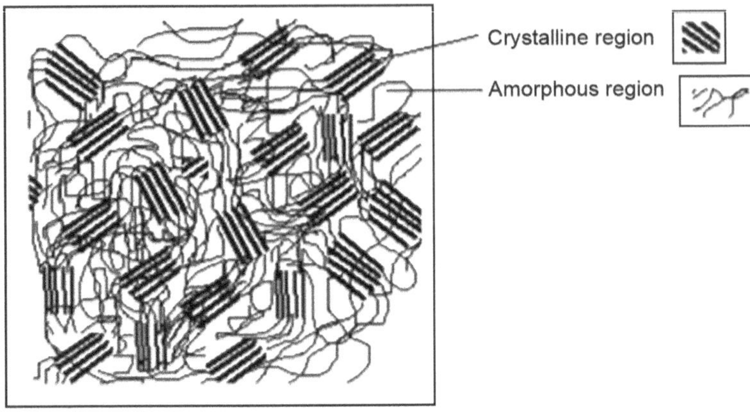

1.8 Crystalline and amorphous region of a polymer.

Most commodity fibers have crystallinity values between 20% and 60%, and crystallinity and molecular orientation together determine the physical properties of the fibers. High strength and high modulus fibers tend to have high crystallinity values, in the range of 90%.

The melting point is the temperature at which a crystallite structure is no longer stable and may be regarded as a measure of the tendency of a polymer to crystallize. The arrangement of the molecules in the structure will play a great role in determining the melting point of a semi-crystalline substance, and therefore once the crystalline structure of large polymeric materials is reached, it is possible to understand the effect of the chemical structure on the melting point. If the polymer does not have any crystallinity, then the fiber does not show a melting transition but only a glass transition temperature (T_g) which corresponds to the change from a brittle glassy to a soft rubbery state.[35,36]

The development of fiber structure is the result of several technical operations including spinning, drawing and heat setting. Each process plays an important role in determining the behavior of the polymer during the next step, thus affecting the final structure and quality of the fiber. Molecular orientation, which is the extension of macromolecules and other structural units along the fiber axis, determines the tensile strength, sorption of solvents, modulus, optical behaviors, etc.[37–39]

1.5 Effects of molecular weight on fiber spinnability, structure and properties

The molecular weight of a polymer has a significant effect on its spinnability. The word does not have a precise meaning in relation to fiber technology, although it often means 'fiber forming,' and the definition includes the physical and chemical aspects of fiber formation from fluid threads without breakage. This means that the range of flow characteristics of the polymer in liquid form is such that no breakage occurs during the spinning process. The spinning process is important because the spinning parameters determine the elongational flow and therefore have a direct effect on the final characteristics of the fibers.[40–43]

Melt spinning has been used for many years in the polymer industry because it provides an effective means of combining two or more materials to produce fibers. Multicomponent fibers are made in this way, and each component material plays an important role in determining the end characteristics of the final product. The molecular weight of each individual material is therefore significant because the chemical and physical interaction between them is partially guided by the available quantities, which are directly linked to the molecular weight. A more complex phenomenon occurs when the different materials form phases during the process of fiber formation. Manipulating the pre-spinning conditions of a polymer or polymer composite can alter the fiber morphology as well as its mechanical and chemical structure. Other factors influencing fiber properties include time, temperature dependent molecular motion, crystallization, length and length-distribution of the molecules, and various intermolecular interactions.[44–47]

In dry spinning, the solution containing the polymer is forced through a fine jet, followed by evaporation of the solvent, which results in the formation of fiber. Dry spinning can therefore be considered a process of structure formation in a bi-component system involving polymer and solvent. This makes analysis of the dry spinning process more complex than that of melt spinning. It is treated as a single component system involving only the polymer because the evaporation rate of the solvent at the boundary between the two phases may not be easily defined. In wet spinning, a solution of polymer is forced through a nozzle into a non-solvent for the polymer. Here, the mass transfer for both the solvent and the polymer is considered. This makes the process analysis even more complex than that of dry spinning.[48–51]

1.6 Flow properties of polymers

The flow characteristics of a polymer largely determine the type of machinery used for extrusion and influence the choice of the process and processing conditions. The rheological and thermal properties of a polymer determine the flow characteristics and are discussed here. The rheological properties define the manner in which a material will deform when certain stresses are applied, a characteristic that is important in the plasticizing zone and the die-forming regions of the process. In a bulk process, the flow properties of the material are directly linked to the transport behaviors, which are determined by the coefficient of friction, the size and shape of the particles, and the density. In polymeric melts, the shape of the molecules will in effect dictate the dynamic behaviors and rheological characteristics. It is therefore of the greatest importance to understand these flow parameters, particularly for the purpose of synthesizing new fiber-forming polymers for industrial and commercial applications.[52–54]

Rheology is generally defined as the study of flow. Polymer rheology is the scientific study of how polymeric fluids deform when subjected to external stresses. When synthesizing a polymer for a particular application, an understanding of the extruder environment in which that polymer will be used is the key to determining the properties of the polymer, so that its response matches the predetermined environment. The chemical and molecular structure of polymer melts will generally have specific effects on the dynamic behaviors and properties of the polymer rheology. It is therefore possible to engineer an end product that fits the predetermined application by manipulating the molecular composition of the polymer. Understanding the molecular structure of polymers can also provide an insight into the visco-elastic properties of polymer melt. This is essentially an understanding of how polymer energy is dissipated in the form of heat during deformation, and how the elastic nature of the polymer will store this energy and use it to restore the polymer to its pre-deformed condition. Such materials are said to have 'memory.'[29,55]

The elastic nature of polymeric materials may be attributed largely to the long chain molecules, which cause entanglements to occur at the molecular level, forming a network through Brownian motion that has a greater effect at elevated temperatures. During deformation, a polymer will alter its molecular configuration and, at the end of the deformation, will return to equilibrium due to Brownian motion.[56,57]

Because of their significance both in theory and practice, the rheological properties of solutions and melts of fiber-forming polymers have been extensively studied over the years. Several research groups and industries have pointed out that varying the viscosity of a fiber-forming polymer through the manipulation of the chemical and production parameters (such as the degree of polymerization compared with the resulting fiber structure) has proved that fiber structure and characteristics are greatly influenced by rheological properties. However, it remains difficult to formulate constitutive characteristic equations that can be precisely used to describe the connection between viscosities, shear stress and shear rate.[54,58]

In melt blowing, for example, surface tension-driven instabilities have been observed both in laboratories and in industrial processes when finer fibers are generated. These create droplets that are dispersed in the fiber mat. Systematic research has proved that altering the rheological properties of the polymer suppresses, and in some cases eliminates, their stability.[59]

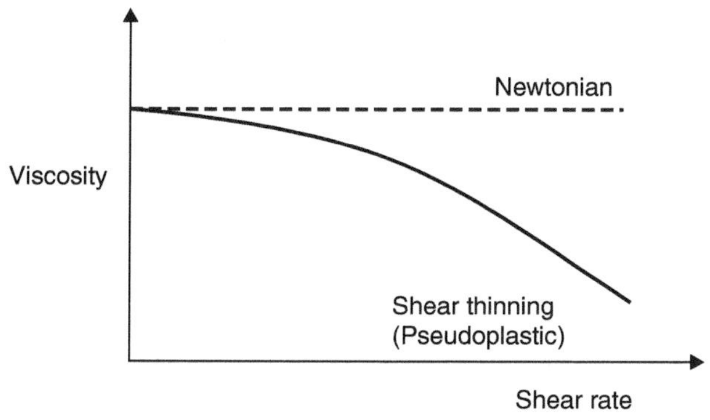

1.9 Shear thinning characteristics of polymer.[60]

Shear thinning behavior, in which viscosity varies inversely with respect to shear rate, is one of the most remarkable properties of polymeric materials. When the shear rate, i.e., the rate of extrusion through a die is increased, the viscosity will decrease because molecular disentanglement and alignments of the polymer chains will occur. Figure 1.9 shows the shear thinning behavior of polymers. The higher the shear rate, the greater the force required to push the polymer melt through a die.[60] The qualities of fibers formed from different types of polymers are determined by the rheological features of the polymer during processing, and the computation of such processes will take into consideration the shear rate to optimize the process. Rheology measurements can be used for characterization and determination of material processability and can also be used as input data for computer simulations. This is because the rheology of a material is more sensitive to the structural aspects of the materials, such as branching and molecular weight tail, and is also a fast method compared to many other options of polymer liquid characterization.

1.7 Principles of solidification during fiber formation

The formation of fibers from both natural and synthetic polymers is achieved through the transformation of a polymer fluid, either molten or dissolved in a solvent, into solid filaments. The chemical process for the formation of different kind of fibers for varied applications differs, but the principle of solidification from liquid to solid remains the same. The rate of solidification of a polymer melt or solution is an important factor which determines the rate of production and the properties of the resulting fibers. It is a phenomenon that has been validated by different research groups in laboratories and industry.[61]

For solidification to occur in a polymer melt, there must be a heat transfer process. This can be done in a controlled environment where a specific shaping and stabilization of the fiber is expected. The rate of solidification is determined by the rate of heat transfer, but the thermal conductivity characteristics of the polymer may generate a larger thermal gradient within the product, causing varied polymer structure within the fiber and significant change in their physical and chemical structure. The choice of the shaping process and equipment used can also affect the solidification process. The rheological properties of the polymer, in influencing the choice of shaping process, will therefore indirectly affect the solidification process. In polymer reactions which occur in solvents, the solidification of the polymer into a desired fiber structure will be determined by the rate of evaporation of the solvents.

1.8 Fiber formation techniques

Rapid advances in the synthetic fiber industry during the 1940s led to an increase in melt spinning technology in order to improve productivity and control the quality of the end products. In melt spinning, fiber diameter may be precisely predicted based on the temperature and tension of the running filaments if the spinning conditions and rheological properties of the polymer are defined and maintained during the spinning process. Such predictions, however, will assume that no crystallization occurs during the spinning process. In cases where crystallization does occur, several parameters must be considered if an accurate mathematical simulation is to be done: the molecular orientation that results from the elongational melt flow, the influence of molecular orientation on the crystallization kinetics, significant change in the rheological properties, and the kinetics of the non-isothermal crystallization. There are several spinning techniques, each tailored to work with different kinds of polymer and/or the desired fiber end products.

1.8.1 Melt spinning

The melt spinning process has been in use since the late 1950s. The polymer melt is extruded and drawn at speeds of 1000–6000 m/min. Lower spinning speeds result in an undrawn yarn, which is then further drawn and heat-treated. At higher spinning speeds, the fiber is partially or fully drawn, and can be texturized, drawn further, or used as spun yarn. PET and nylons are examples of such polymeric fibers processed using melt spinning. The resulting fiber structure is a function of the spinning speed, the drawing mechanism, heat setting conditions etc. The stress–strain behavior of the material is also a function of the spinning speed, because the polymer chains orient and crystallize and the fiber structure develops as the spinning speed increases. Reproducible experimental results by several research groups have shown that the molecular orientation improves much more than crystallinity with the spinning speed. However, the crystallite size will increase with spinning speed.[62]

In high speed spinning, the spinning tension is a function of inertial force and air resistance. However, it is generally agreed that the structural formation is defined mainly by inertial force and that air resistance may be ignored in speeds over 8000 m/min. But researchers have also reported that air resistance could influence the formation of the structure of fine filaments at spinning speeds between 4000 and 5000 m/min. The mechanical properties of the resultant fibers can be improved by heat-treatment of fibers in the spinline. Fibers have also demonstrated improved properties when cooled in the spinline instead of undergoing heat-treatment. Quenching immediately after necking results in highly oriented amorphous yarns. Cooling can also give precise control of the temperature of the cooling agent. In general, ultra-high-speed spinning has the potential to produce fibers with improved performance properties, and further development is needed to achieve high efficiency and high quality fibers.

1.8.2 Solution spinning

Significant developments have recently been made in the fiber solution spinning process due to increased understanding of theoretical spinning technology, especially in acrylic fibers. Solution spinning technology is primarily based on spinning a solution or suspension that precipitates or coagulates to form a solid fiber as an end product. Solution spinning is categorized into two methods: dry spinning and wet spinning. In dry spinning, the polymeric solution solidifies when evaporation of the solvent occurs. Wet spinning can be further divided into three classifications, based on physiochemical principles. In the liquid crystal process, lyotropic polymer liquid crystalline solution is solidified through the formation of a solid crystalline region in a solution. The gelation mechanism solidifies the polymer solution through formation of intermolecular bonds in the solutions. In the phase separation process, polymer rich and polymer lean phases are part of the solution.[63]

1.8.3 Highly aesthetic fiber spinning

Fibers with unique sensual characteristics, such as color, appearance, feel, bulkiness and texture, are commonly referred to as aesthetic fibers. Aesthetic fibers are highly fashionable and expensive, and their production process and advanced technology are also costly. Materials that may require specialized finishing are processed in this manner. For example, the addition of inorganic materials in the mixed-spinning process increases the specific gravity of a fiber and improves the drapability of the fabrics. This can be varied by weight percentage of the additive, depending on the desired quality. Addition of these organic polymers or a composite may be done between the polymerization and spinning processes.[64]

1.8.4 Anisotropic fiber spinning

High strength fibers made from anisotropic polymers are more difficult to synthesize because of the technical process required to form an anisotropic solution from the solid polymer. Aromatic polymers such as polyimides and heterocyclics, for example, are characterized by extended chain conformation and are known to form anisotropic solutions or melts that have liquid crystal behavior at the correct solution temperature and composition. During the fiber formation process, the liquid crystal domain in the polymer melt will undergo orientation as a result of the influence from the shear and elongation flow. When cooled, these anisotropic polymers also exhibit phase transition, forming highly crystalline solids. Fiber spinning of anisotropic polymers will therefore result in highly crystalline and oriented fibers with high strength and modulus. Anisotropic polymers are classified into two types: lyotropic, in which solutions respond to different solvent types and concentrations in a varying manner, and thermotropic polymers, in which melts respond to temperature changes by behaving like liquid crystals. In a molten or solution state, anisotropic polymers display mesomorphic behavior similar to low molecular weight liquids, but differ in having have higher molecular weights.[65]

1.8.5 Thermotropic liquid-liquid crystal polymer spinning

Fibers with very high strength and modulus can be fabricated from polymers that have a molecular structure in which the chains are packed in small cross-sectional areas with strong bonds and low elongation. However, some of these aromatic polymers have melting points that are higher than their decomposition temperatures due to the rigidity of their molecules. It is therefore impossible to process them in thermotropic liquid crystal form.

It has been shown that this can be overcome by a slow introduction of flexible alkyl groups into the main chain, by introducing substituents into the individual aromatic ring in the main chain, by co-polymerization of more than one rigid molecule, or by the introduction of non-molecular structures.

Combinations of these techniques have been shown to provide a means of creating thermotropic LCPs that can be injection-molded or melt-spun. The chemical structures of thermotropic LCP fibers are varied and based on the original polymer material characteristics and conditions. The choice of synthesis process is therefore the key to achieving the desired end product characteristics and qualities. The current challenge, however, in the spinning of thermotropic LCPs is to improve cost effectiveness through performance process optimization.

1.8.6 Gel spinning

Hiroshi *et al.* have defined the gel spinning process as a method of achieving high strength fibers through an intermediate gel-like state. In gel spinning, the most important feature of the polymer is the degree of polymerization. Polymers with an average molecular weight above 600 000 g/mol are effectively processed by this technique. Although gel spinning is a relatively long-standing process, in its current adaptation, polymer of an extremely high molecular weight is dissolved in a solvent at low concentrations of about 1–2 percentage of polymer by weight, making a highly viscous solution. The solution is then dry- or wet-spun to a fiber that retains most of the solvent and is actually a gel of polymer and solvent.[66,67] The gel spun fiber is further drawn with the removal of the solvent to produce a very highly oriented and ordered structure with high strength and modulus.

1.8.7 Spinning of ultra-fine fibers and optical fibers

Any fiber that is less than 0.7 denier is referred to as ultra-fine fiber. These fibers can be processed by significantly reducing the polymer output at the spinneret while drawing it with a large draw ratio. Currently, ultra-thin continuous filaments may be produced by various methods including direct spinning, conjugate spinning and multi-layer type spinning. Ultra-fine fibers can be produced by methods that include melt blowing, electrospinning, flash spinning, polymer-blend spinning, and centrifugal spinning. Softness, flexibility, smoothness, fine textile structures, micro-pockets in fibers, quick stress relief, low resistance to bending and many other properties characterize ultra-fine fibers. These qualities are definitive in determining combinatorial processing techniques.[68–70]

An optical fiber is composed of a core made from a light transmitting material which is transparent and has a very low refractive index. Depending on the application of such fibers, the choice of the spinning method is based on the polymer material used.[71]

1.9 Post-fiber formation treatments

The reasons for post-fiber formation treatments are many and varied. Fiber treatment may involve drawing, twisting, texturing, coating, further thermal treatment, chemical, or combinatorial processes. Growing environmental concerns may require fibers to be treated to minimize toxic emissions. Fibers for some applications may require reinforcement, which can be achieved by chemical or thermal treatments. The need to prolong the lifetime of fibers for certain applications will also call for post formation treatments. Fibers designed for biological implants applications, for example, should be treated to minimize side effect reactions with the body, and its lifetime should be such that it can remain functional during the life of the person or animal. Other processes can be determined and influenced by aesthetic modification for desired qualities.

1.10 Current trends

The consumer market-driven demand for improved products has led to a growth in production techniques for a variety of materials. There is an increase in the applications of polymeric fibers, the limited capabilities of naturally occurring fibers having caused a growth in research into man-made fibers which can be engineered for specific applications. The reliability of polymeric fibers is influencing applications in several areas, including the biomedical, aviation, and automobile industries. The need for materials with nanocomposite structures and high performance standards has driven research in both laboratory and industry. Another area that has attracted attention in the past decade and remains of interest is that of submicron and nanofibers. Several polymers and processing techniques continue to be investigated.

1.11 Conclusion

The demand for man-made polymers and polymeric fibers is growing due to the flexibility of processing, which allows a material with specific mechanical, electrical, optical, and chemical features to be tailored to fit a specific application. This has led to an increase in research to find more source polymers. However, the growth in man-made polymers has raised environmental and health issues. Research and developmental activity has increased in the development of biopolymers to meet the current and future demand for sustainability. There is also a need to develop processing equipment for the production of fibers in the nanometer range. The future of polymer production appears to be interesting and commercially sustainable.

1.12 References

1. Goodman, I. (1968), *Synthetic Fibre-Forming Polymers*, London: Royal Institute of Chemistry. 4, 79 p.
2. Salamone, J.C. (1999), *Concise Polymeric Materials Encyclopedia*, Boca Raton: CRC Press. 1706 p.
3. Grineviciute, D. Stankute, R. and Gutauskas, M. (2010), Influence of new fibre-forming polymers structure on garment hand parameters. *Materials Science-Medziagotyra*. **16**(2): pp. 144–147.
4. Shepelev, A.V. (2009), Physical chemistry of fibre-forming polymers transformation of emission brightness in polymer fibres. *Fibre Chemistry*. **41**(6): pp. 377–378.
5. Druzhinina, T.V., Abronin, I.A. and Bikkullova, A.R. (2006), Quantum-chemical calculation of the energy characteristics of graft polymerization of methyl methacrylate to fibre-forming polymers. *Fibre Chemistry*. **38**(3): pp. 196–199.
6. Peacock, A.J. and Calhoun, A.R. (2006), *Polymer Chemistry : Properties and Applications*, Munich, Cincinnati, Ohio: Hanser Gardner Publications. xix, 397 p.
7. Anon (2010), Polymer science and technology editorial. *Polimeros-Ciencia E Tecnologia*. **20**: pp. E1–E1.
8. Mikheev, Y.A., Guseva, L.N. and Zaikov, G.E. (1997), Characteristic features of heterophase macromolecular chain reactions involving antioxidants in a non-crystalline polymer matrix. *Uspekhi Khimii*. **66**(1): pp. 3–30.
9. Jing, J-.H., Huang Y.-D. and Liu, L. (2011), Formation of fibre-bundle/matrix interlayer in carbon/carbon composites. *Journal of Inorganic Materials*. **26**(12): pp. 1309–1313.
10. Baker, D.A., Gallego, N.C. and Baker, F.S. (2012), On the characterization and spinning of an organic-purified lignin toward the manufacture of low-cost carbon fibre. *Journal of Applied Polymer Science*. **124**(1): pp. 227–234.
11. Hearle, J.W.S. and Textile Institute (Manchester England) (2001), *High-Performance Fibres*, Boca Raton Cambridge, England: CRC Press, Woodhead Publishing Limited. xi, 329 p. 12. Polymers, C.o.L.C. and Board, N.M.A. (1990), *Liquid Crystalline Polymers*, Atlanta, GA: The National Academic Press.
13. Muth, M.A., Carrasco-Orozco, M. and Thelakkat, M. (2011), Liquid-crystalline perylene diester polymers with tunable charge-carrier mobility. *Advanced Functional Materials*. **21**(23): pp. 4510–4518.
14. Goto, T., Maeda, M. and Hibi, S. (1989), High-strength and high toughness aromatic polyamide fibre. I. Preparation and properties of block poly(para-phenylene and 4,4′-diphenyl ether terephthalamide) fibre. *Journal of Applied Polymer Science*. **37**(4): pp. 867–875.
15. Preston, J. (1981), High-strength high modulus fibres from aromatic polymers. *Abstracts of Papers of the American Chemical Society*. **181**(Mar): pp. 81–CHED.
16. Riedel, U. and Nickel, J. (1999), Natural fibre-reinforced biopolymers as construction materials – new discoveries. *Angewandte Makromolekulare Chemie*. **272**: pp. 34–40.
17. Mohanty, A.K., Misra, M. and Drzal, L.T. (2005), *Natural Fibres, Biopolymers, and Biocomposites*, Boca Raton, FL: Taylor and Francis. 875 p.

18. Klemm, D. Heublein, B. and Fink H.P. (2005), Cellulose: Fascinating biopolymer and sustainable raw material. *Angewandte Chemie-International Edition.* **44**(22): pp. 3358–3393.
19. Agamuthu, P. and Faizura, P.N. (2005), Biodegradability of degradable plastic waste. *Waste Management and Research.* **23**(2): pp. 95–100.
20. Tokiwa, Y. Calabia B.P. and Ugwu, C.U. (2009), Biodegradability of plastics. *International Journal of Molecular Sciences.* **10**(9): pp. 3722–3742.
21. Hoidy, W.H., Al-Mulla, E.A.J. and Al-Janabi, K.W. (2010), Mechanical and thermal properties of PLLA/PCL modified clay nanocomposites. *Journal of Polymers and the Environment.* **18**(4): pp. 608–616.
22. Berry, G.C. (1983), Principles of polymer systems – Rodriguez, F. *Journal of the American Chemical Society.* **105**(11): pp. 3746–3746.
23. Sandler, S.R. and Karo, W. (1974), *Polymer Syntheses. Organic Chemistry A Series of Monographs*, New York: Academic Press.
24. Bartke, M. and Diener, A. (2012), Reactor for solvent-free polymer processing at CPP Chemical Plants and Processes [http://www.cpp-net.com/process_engineering/-/article/32218883/37074576/Reactor-for-solvent-free-polymer-processing/art_co_INSTANCE_0000/maximized/]
25. Albright, L.F. (1985), *Processes for Major Addition-Type Plastics and Their Monomers*, Huntington, N.Y.: R.E. Krieger Pub. Co.
26. Halverson, J.D. Grest G.S. and Grosberg, A.Y. (2012), Rheology of ring polymer melts: from linear contaminants to ring-linear blends. *Physical Review Letters.* **108**(3): 038301.
27. Ahn, S.K., Deshmukh, P. and Kasi, R.M. (2010), Shape memory behavior of side-chain liquid crystalline polymer networks triggered by dual transition temperatures. *Macromolecules.* **43**(17): pp. 7330–7340.
28. Couchman, P.R. and Karasz, F.E. (1977), Some domain-size effects in polymer transitions. *Journal of Polymer Science Part C-Polymer Symposium.* **63**(1): pp. 271–280.
29. Peiffer, D.G. (1978), Degrees of freedom of polymeric molecules and their effect on physical-properties. *Journal of Macromolecular Science-Physics.* **B15**(4): pp. 595–611.
30. Kim, A. and Musfeldt, J.L. (1998), Understanding chemical structure physical property relationships in polymers through molecular modeling and thermal analysis techniques. *Journal of Chemical Education.* **75**(7): pp. 893–896.
31. Hill, R. (1949), Some aspects of polymer constitution and fibre properties. *Chemistry and Industry.* (27): pp. 443–443.
32. Hill, R. and Walker, E.E. (1948), Polymer constitution and fibre properties. *Journal of Polymer Science.* **3**(5): pp. 609–630.
33. Mandelkern, L. (1981), Polymer principles in the teaching of undergraduate physical-chemistry. *Abstracts of Papers of the American Chemical Society.* **181**(MAR): pp. 10–CHED.
34. Flory, P.J. (1953), *Principles of Polymer Chemistry. The George Fisher Baker non-resident lectureship in chemistry at Cornell University*, Ithaca: Cornell University Press. 672 p.
35. Ball, P. (1995), Polymer science. Gourmet macromolecules. *Nature.* **377**(6545): pp. 101–102.
36. Cheng, S.Z. (2007), Materials science: polymer crystals downsized. *Nature.* **448**(7157): pp. 1006–1007.

37. Marcincin, A. (2002), Modification of fibre-forming polymers by additives. *Progress in Polymer Science*. **27**(5): pp. 853–913.
38. Jana, R.N., Kumar, A. and Im, C. (2009), Molecular orientation of polyurethane based liquid crystal polymers by corona poling. *Journal of Macromolecular Science Part a-Pure and Applied Chemistry*. **46**(10): pp. 1001–1006.
39. Patel, P.D. and Bogue, D.C. (1981), The effect of molecular-orientation on the mechanical-properties of fibre-filled amorphous polymers. *Journal of Rheology*. **25**(4): pp. 461–462.
40. Tsai, J.S. and Lin, C.H. (1991), The effect of molecular-weight on the cross-section and properties of polyacrylonitrile precursor and resulting carbon-fibre. *Journal of Applied Polymer Science*. **42**(11): pp. 3045–3050.
41. Tan, L.J., Chen, H. and Pan, D. (2008), Investigating the spinnability in the dry-jet wet spinning of PAN precursor fibre. *Journal of Applied Polymer Science*. **110**(4): pp. 1997–2000.
42. Dutta, A. and Nadkarni, V.M. (1984), *Identifying critical process variables in poly(ethylene-terephthalate) melt spinning. Textile Research Journal*. **54**(1): pp. 35–42.
43. Samuels, R.J. (1974), *Structured Polymer Properties: The Identification, Interpretation, and Application of Crystalline Polymer Structure*, New York: Wiley. xi, 251 p.
44. Zeng, J., Chen, J. and Wang, Y. (2008), Development of relatively selective, chemically and mechanically robust solid-phase microextraction fibres based on methacrylic acid-trimethylolpropane trimethacrylate co-polymers. *Journal of Chromatography. A*. **1208**(1–2): pp. 34–41.
45. Buck, M.E. and Lynn, D.M. (2010), Functionalization of fibres using azlactone-containing polymers: layer-by-layer fabrication of reactive thin films on the surfaces of hair and cellulose-based materials. *ACS Applied Materials and Interfaces*. **2**(5): pp. 1421–1429.
46. Salem, D. (2001), Future view of structure formation in polymeric fibres. *Sen-I Gakkaishi*. **57**(1): pp. P31–P31.
47. Baltussen, J.J.M. (2005), Polymeric and inorganic fibres. *Advances in Polymer Science*, Berlin, New York: Springer. vi, 168 p.
48. Wu, Q., Chen, N. and Li, L. (2012), Structure evolution of melt-spun poly(vinyl alcohol) fibres during hot-drawing. *Journal of Applied Polymer Science*. **124**(1): pp. 421–428.
49. Monobe, K. (April 1979), Fundamental approach for fibre formation of crystalline polymers. *Abstracts of Papers of the American Chemical Society*, pp. 124–124.
50. Lotze, G., Stephani, G., Löser, W. and Fieldler, H. (1991), Fundamentals of fibre formation during melt extraction. *Materials Science and Engineering a-Structural Materials Properties Microstructure and Processing*. **133**: pp. 680–683.
51. Remmler, T. (1999), Flow properties of polymer melts. *Kunststoffe-Plast Europe*. **89**(5): pp. 126–128.
52. Acierno, D., Bianchi, E., Ciferri, A., Tealdi, A., Torre, R. and Valenti, B. (1976), Bulk properties of synthetic polymer-inorganic salt systems .3. Flow behavior and glass-transition of salted polycaproamide. *Journal of Polymer Science Part C-Polymer Symposium*. (54): pp. 259–269.
53. Horio, M., Fujii, T. and Onogi, S. (1964), Rheological properties of polyethylene melts – effects of temperature + blending. *Journal of Physical Chemistry*. **68**(4): pp. 778–783.

54. Jabbarzadeh, A., Atkinson, J.D. and Tanner, R.I. (2003) Effect of molecular shape on rheological properties in molecular dynamics simulation of star, H, comb, and linear polymer melts. *Macromolecules*. **36**(13): pp. 5020–5031.
55. Shenoy, A.V. and Saini, D.R. (1996), Thermoplastic melt rheology and processing. *Plastics Engineering*, New York: Marcel Dekker. ix, 459 p.
56. Sato, N., Sugiura, K and Ito, S. (1997), Molecular motion in polymer monolayers at the air/water interface. A time-resolved study of fluorescence depolarization. *Langmuir*. **13**(21): pp. 5685–5690.
57. Cerf, R. (1977), Structure dependent hydrodynamic properties and Brownian-motion of polymeric molecules. *Journal De Physique*. **38**(4): pp. 357–376.
58. Lelievre, J.C. (1987), *An introduction to rheology. Biorheology*. **24**(4): pp. 415–415.
59. Gillgren, T., Alven, T. and Stading, M. (2010), Impact of melt rheology on zein foam properties. *Journal of Materials Science*. **45**(21): pp. 5762–5768.
60. Dupuis, D. (2009), Textile processes: the importance of rheology. *Mecanique and Industries*. **10**(1): pp. 21–25.
61. Mercader, C., Lucas, A. and Derre, A. (2010), Kinetics of fibre solidification. Proceedings of the National Academy of Sciences of the United States of America. **107**(43): pp. 18331–18335.
62. Liang, J.-Z., Zhong, L. and Wang, K. (2012), Analysis of melt spinning master-curves of low density polyethylene. *Journal of Applied Polymer Science*. **125**(3): pp. 2202–2206.
63. Zhang, X.F., Min, B.G. and Kumar, S. (2003), Solution spinning and characterization of poly(vinyl alcohol)/soybean protein blend fibres. *Journal of Applied Polymer Science*. **90**(3): pp. 716–721.
64. Wallace, G.G., Campbell, T.E. and Innis, P.C. (2007), Putting function into fashion: Organic conducting polymer fibres and textiles. *Fibres and Polymers*. **8**(2): pp. 135–142.
65. Calvert, P. (1987), Introduction to physical polymer science – Sperling, LH. *Nature*. **326**(6109): pp. 216–216.
66. Ohta, Y. (2004), Structural development of ultra-high strength polyethylene fibre during gel-spinning process. *Sen-I Gakkaishi*. **60**(9): pp. P451–P455.
67. Yi, D.K., Yoo, S.J. and Kim, D.Y. (2002), Spin-on-based fabrication of titania nanowires using a sol-gel process. *Nano Letters*. **2**(10): pp. 1101–1104.
68. Igoshi, T., Tomisawa, N. and Hori, Y. (2011), Polyester polymer alloy as a high-performance membrane, in A.K.H.Y.A.C.M.M. Saito, Ed., *High-Performance Membrane Dialyzers*,. pp. 148–155.
69. Narisawa, I., Kuriyama, T. and Schemer, B. (1996), Design and control of morphology for high performance polymer alloys. *Macromolecular Symposia*. **101**: pp. 273–280.
70. Takeichi, T., Saito, Y. and Agag, T. (2008), High-performance polymer alloys of polybenzoxazine and bismaleimide. *Polymer*. **49**(5): pp. 1173–1179.
71. Assemat, E., Dargent, D. and Picozzi, A. (2011), Polarization control in spun and telecommunication optical fibres. *Optics Letters*. **36**(20): pp. 4038–4040.

2
Understanding the behaviour of synthetic polymer fibres during spinning

E. DEVAUX, Ecole Nationale Supérieure des Arts et Industries Textiles (ENSAIT), and Centre Européen des Textiles Innovants (CETI), France

DOI: 10.1533/9780857099174.1.31

Abstract: The spinning of synthetic polymers is a field of plasturgy, where the main objective is to process materials with a particular directionality. This macromolecular orientation determines the end products' functional characteristics and mechanical behaviour. Thermoplastic macromolecules have a natural tendency towards relaxation and isotropic spherulitic crystallization. As such, it is necessary to adapt the spinning parameters with the aim of maximizing chain orientation. The principal problem of spinning is adapting the polymer viscosity in the molten state or in solution in order to disentangle the macromolecular chains. As the viscosity depends directly on the molecular weight of the material, it is necessary to find an adequate compromise between the desired mechanical behaviour and the rheological properties of the polymer. Most thermoplastic polymers have a strong tendency to crystallize during cooling. This phenomenon generally results in an isotropic arrangement of the macromolecules and must be controlled to ensure orientation in the fibre direction. The relaxation of the macromolecular chains at the exit of the dies causes a die swell phenomenon, which must also be treated to optimize the final properties of the yarns.

Key words: thermoplastic polymer rheology, crystallization, macromolecular superstructures, extrusion, die swell, drawing.

2.1 Introduction

Since the discovery of the qualities of natural silk around 3000 BC, mankind has tried to imitate nature by producing synthetic materials with similar properties to this material. The exceptional mechanical and handling properties of natural silk were and remain an absolute reference in the field of fibres. The 'artificial silk' patented in 1884 by French Bernigaudde Chardonnet is often seen as the first significant development in man-made fibres. With the appearance of plastics at the beginning of the twentieth century, and the studies undertaken on the macromolecular structure of polymers by

Hermann Staudinger, there has been an increasing development of these materials for numerous applications, including textiles. Polyamide 6.6 fibre (nylon), developed by Du Pont de Nemours in 1938, was a revolution in clothing and military uses. This fibre, the functional characteristics of which are rather close to silk, has exceptional tenacity, flexibility and lightness. With the progress of macromolecular chemistry at that time, it was the first of a long series of diverse synthetic fibres.

With rare exceptions, man-made fibres are obtained from thermoplastic polymers. These materials, generally formed as a result of the condensation of monomers, are made from long linear or slightly ramified macromolecular chains. These configurations give to polymeric material plasticity at relatively low temperatures, making it possible to process it in a simple manner. Thermoplastic polymers have seen significant development in recent decades because the property of liquefying at temperature yields the potential for recycling.

In a first approximation, the processing of synthetic fibres consists in bringing the polymer to a liquid state (by melting or solubilization in a suitable solvent), to pass it through an adapted die, and then to stretch it while it is in the rubbery state. The material is finally cooled, or the solvent is eliminated, before the fibres are wound. The objective of drawing is to optimize the orientation of the macromolecules in the direction of the material flow, which in turn optimizes the final properties of the yarn. From the point of view of the fundamental characteristics of polymers, it is the viscosity and rheology (in the liquid or solid state), which control the workability of the fibre.

2.2 Molten state rheology of thermoplastic polymers

For any processing technology of a thermoplastic polymeric material (injection moulding, calendering, extrusion-blowing, compression moulding, spinning, etc.), it is the material flow which defines the suitability of the method considered. For simple fluids, viscosity remains the most important parameter in understanding the behaviour of the liquid. The rheology of polymers is, however, more complex because of their non-ideal behaviours compared to conventional liquids. Besides viscosity, molten polymers, or polymers in solution, also have elasticity and relaxation characteristics which strongly influence their behaviour in the processing tools.

The rheological properties of a polymer depend mainly on chemical structure and molecular weight, as well as temperature. The possible presence of additives and their concentration is also a key parameter relating to the rheological behaviour of a polymer. Viscosity η (expressed in Pa.s) for an ideal liquid is independent of the shear rate. During flow, such a liquid follows Newton's law, which expresses the relationship between the shear stress γ and the gradient rate of deformation $\frac{d\gamma}{dt}$.

$$\sigma = \eta \frac{d\gamma}{dt} \quad [2.1]$$

Experimentally, Newton's law is generally followed only by gases and by simple liquids in laminar flow. It is only at a very low deformation rate and shear stress that solutions of macromolecules, molten polymers, emulsions and suspensions, display approximately Newtonian behaviour.

At higher shear rates, three types of deviations are observable when compared to ideal Newtonian flow (see Fig. 2.1).The first kind of deviation relates to the existence of a flow threshold (yield point). In the case of a Bingham fluid, flow occurs only when the yield stress is exceeded. The second type of deviation is shear thickening, observed where the viscosity increases with shear rate. This is the case for a dilating fluid, behaviour which is seldom apparent in polymers. Last, where viscosity decreases with increase in shear rate, fluxing is observed and such fluids are usually referred to as pseudoplastic fluids. This last phenomenon is a general characteristic of thermoplastic polymers. Flow effects may also be time dependent. Where viscosity does not depend only on the shear rate, but also on the duration of the applied stress, fluids are thixotropic. Polymers in a molten state thus behave as pseudoplastic fluids having thixotropic characteristics.

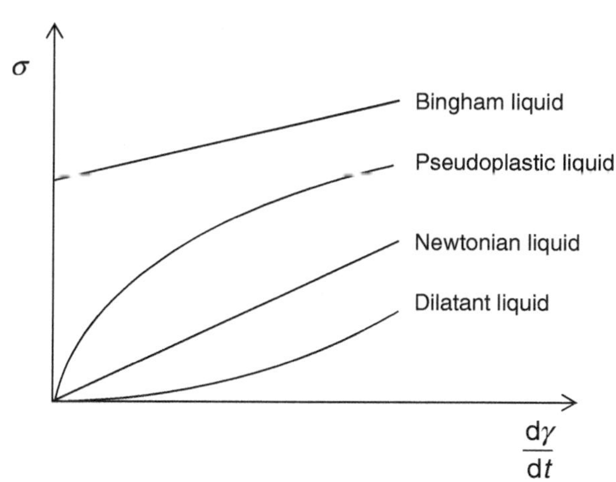

2.1 Different kinds of deviations compared to the Newtonian flow as a function of the shearing rate.

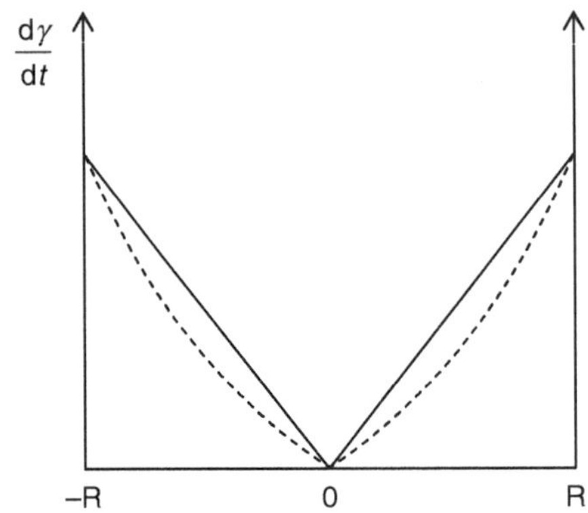

2.2 Shearing rate profiles in a spinning capillary.

At increased mechanical stress, molten plastics are characterized by a decrease in viscosity. This reduction in viscosity when $\frac{d\gamma}{dt}$ increases can be explained only by the fact that increasing mechanical stresses gradually destroy the structured organization of the molten polymer. In applications such as spinning, where the liquid polymer is constrained to cross a die, the shear rate is not constant but varies in the capillary. There is thus a maximum shear rate localized at the walls of the holes, while at the centre of the capillary, it is virtually non-existent (see Fig. 2.2).

Polymer viscosity is affected by several factors, such as temperature, pressure, shear rate, molecular weight and the presence of additives. Flow occurs in the molten state when the macromolecular chains slip over each other. The ease of this movement depends on the mobility of the polymeric chains, the intermolecular forces and the sequences ensuring the material cohesion. For a constant shear rate, viscosity decreases with an increase of temperature, and this effect increases with increasing shear rate. In most transformation processes, the flow of polymers in the molten state results from intense shearing; thus the dependencies of temperature and apparent viscosity have considerable practical importance. Newtonian viscosity, an intrinsic rheological parameter of the material, is related to macromolecular characteristics such as the glass transition temperature, average molecular weight, weight distribution, and the connection of side chains. The energy of activation for a polymer to flow in the liquid state increases when the size of the lateral groups increases, and/or when the macromolecular chains are more rigid (see Table 2.1). For very low shear rates, the macromolecular entanglements have time to slip and be destroyed before they impact the orientation of the macromolecules. For higher shear rates, however, the segments between entanglements are oriented before the entanglements disappear. At very high shear rates, the amount of entanglement becomes

Table 2.1 Energy of activation to promote a polymer flowing in the liquid state

Polymer	Energy of activation E (kJ/g.mol)
Polyethylene (high density)	6.3–7.0
Polyethylene (low density)	11.7
Polypropylene	9.0–10.0
Polyethylene terephthalate	19.0
Polystyrene	25.0
Polycarbonate	26.0–30.0

negligible and apparent viscosity reaches a relatively low value, independent of the shear rate.

Taking into account the very strong shear rates associated with the spinning of a polymer through dies, understanding of the rheological behaviour of the material to control the processing parameters is of critical importance. As each polymer has its own rheological characteristics, it is necessary to finely adjust the processing parameters to optimize the final properties of the yarn. Shear rates, which have a direct impact on the production speed of the yarns, are also extremely important since they appreciably modify the viscoelastic properties of the liquid polymer.

2.3 Polymer behaviour during cooling and resulting fibre structure

With rare exceptions, thermoplastic polymers are electrically and thermally insulating. This has a direct impact on the processing parameters and the dimensional stability of the end product. During spinning, the material very quickly passes from a liquid to a solid state (by cooling, coagulation or solvent evaporation), and the control of the evacuation of calories is very important for the final product quality. According to the chemical nature of the polymer considered, there may be a natural tendency to crystallize, i.e. to organize constitutive macromolecules. The crystallization behaviour of a polymer depends on many factors and particularly on the thermomechanical processing conditions. Unlike small molecule crystals, where the structural variety is mainly located at the level of the elementary lattice, polymers have four levels of organization: configuration of the chain, chain conformation and organization of the chains in the elementary lattice, lamellar structure related to chains folding, and organization of these lamellar structures in larger entities (spherulites, for example).

Configuration is definitively determined by the arrangement of the chemical groups, and can be modified for a given polymer only by breakage of the chemical bonds. The conformation of the chains is controlled by interactions

2.3 Spherulitic superstructure in a polypropylene film.

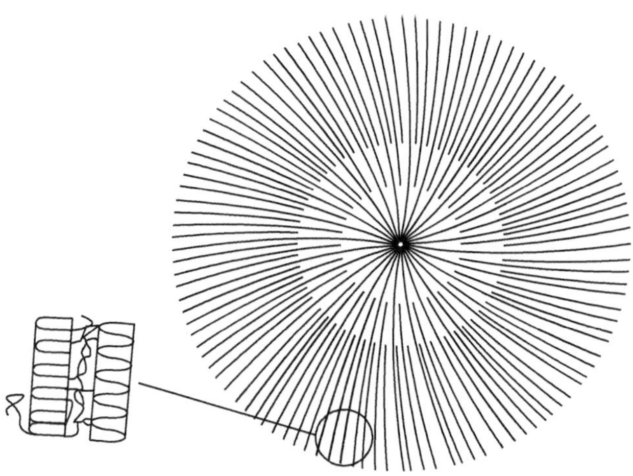

2.4 Spatial distribution of crystalline lamellae in a spherulite.

at short distance between close non-dependent atoms. The regular repetition of these conformations involves the formation of helicoidal structures. From 10 to 30 nm, crystalline lamellae with low thickness are alternated with the residual amorphous phase. At larger scales, the lamellae are organized at long distance to yield morphologies such as spherulites (see Figs 2.3 and 2.4).

During the cooling of a semi-crystalline thermoplastic polymer, the macromolecules are organized around a germ of nucleation. At a particular temperature of thermodynamic balance between crystal and liquid (Tm°), the molar free enthalpies are equal in the solid and liquid phases. Crystallization is theoretically possible as soon as the solid free enthalpy becomes lower than that of the liquid phase, i.e. below the melting point of the material. Organized zones called nuclei are created, with a characteristic variation of free enthalpy ΔG. For a critical volume of nuclei, the probability of growth becomes higher than that of reduction. This occurs where germs are supercritical, or active. Germination is homogeneous when the active germs appear spontaneously in the super cooled liquid phase, but is heterogeneous when the cores are formed on the surface of foreign particles. In the latter case, it is the presence of nucleating agents incorporated into the polymer, or the introduction of functional additives, which cause the nucleation. Crystallization itself is a natural evolution of germination (Lauritzen and Hoffman, 1960; Frank and Tosi, 1961; Gornick and Hoffman, 1966).

In standard conditions of cooling, crystalline growth takes place in the shape of spherulites. This kind of superstructure has fundamental isotropic properties because the growth takes place in three dimensions (see Fig. 2.4), whereas lamellar structures grow in a radial manner away from the central core. In certain cases (in particular for polyethylene), the crystalline lamellae can turn around their growth axis, according to the macromolecular architecture and to the crystallization conditions (see Fig. 2.5) (Keller and Kolnaar, 1997). The spherulites size is related to the number of nucleation sites existing in the material, which mainly depends on the cooling kinetics, the presence of nucleating agents, and on the type of germination. In a spinning application, taking this behaviour into account is very important because the objective is to obtain a strongly anisotropic structure in order to optimize the mechanical properties of the yarn. The orientation of an isotropic polymer in the solid state consists in aligning the polymer macromolecules parallel to the direction of the main mechanical stress, leading to a material with a higher degree of anisotropy due to the chain orientation (Ajji and Dumoulin, 2006). It is well established that the mechanical behaviour of the oriented material, and particularly the strength and stiffness, increases in the molecular orientation direction (Amornsakchai and Songtipya, 2002; Francesco and Duckett, 2004). In most cases, the classical model of drawing proposed by Peterlin (Peterlin, 1971) for crystalline polymers is considered valid (Ziabicki *et al.*, 2004; Schoene *et al.*, 2005).

During the spinning process, the polymer is subjected to an intense flow and to strong shear stresses, which can appreciably modify crystallization behaviour. The various studies available show an acceleration of crystallization under flow (Somani *et al.*, 2000). During shear tests carried out before the isothermal crystallization of polypropylene, Lagasse and Maxwell (1976) highlighted the existence of a critical shear rate, beyond which the induction time decreases when the shear rate increases, the time before crystallization decreases under these conditions. However, certain studies (Vleeshouwers and Meijer, 1996) show that the use of the induction time as a parameter for the measurement of crystallization acceleration is not necessarily relevant.

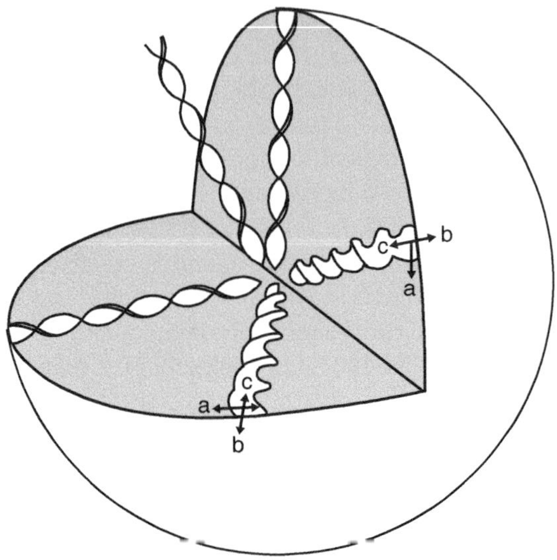

2.5 Representation of crystalline lamellae in polyethylene (Keller and Kolnaar, 1997).

Considering the work on various polypropylenes, it is observed that the polymer reaches its maximum crystallization rate more quickly where the induction time is longest.

Increase of nuclei numbers is one of the major consequences of crystallization below the yield stress. The zones subjected to the greatest shear forces in injected products, for instance, have morphological structures of smaller size than those of the other zones (Kalay and Bevis, 1997). When the stresses are very low and do not promote a sufficient stretching of the macromolecular chains, there can be a macroscopic orientation due to the alignment of the lamellae plans in the direction of the flow. The largest dimension of the lamellae is also aligned to the flow direction. Where the shear rate is increased, the deformation involves an anisotropic growth of the lamellae. The chain elongation in the melt favours the growth of the lamellae in the direction perpendicular to the flow, giving rise to spherulites with ellipsoidal shapes, whose main axes are perpendicular to the direction of the flow.

In recent years, numerous studies have dealt with the incorporation of nanofillers in polymeric materials to improve certain functional properties. In most of these cases, mechanical properties improve significantly. The addition of nanoparticles in a polymer promotes spherulitic crystallization, which may favour the isotropy of the material. A subsequent drawing is then generally necessary to give a satisfactory orientation of the crystalline lamellae (Fujimori *et al.*, 2008; Solarski *et al.*, 2008; Vargas *et al.*, 2010; Murariu *et al.*, 2011).

In the case of very high yield stress, as in the spinning process, morphology is of the fibril type. If the stresses decrease, there will be less fibrillae and they will be more distant from each other. The polymer in between these fibrillae can then crystallize, while taking the fibrillae as germs, and forming parallel lamellae between them. This kind of structure, known as a 'shish-kebab', presents a molecular orientation aligned to the direction of the flow. Nagasawa *et al.* (1973) proposed a model whereby the growth of the lamellae begins with an epitaxial alignment of the crystalline axis compared to the axis of the core fibrillae. Hosier *et al.* (1995) observed an increase in the diameter of the shish-kebabs and a reduction in the longitudinal distance between the lamellae with an increase in the deformation rate. At deformation rates similar to those observed during spinning, the lamellae are plane and normal to the fibrillae, whereas at lower rates, they are inclined and have an S or C shape perpendicular to the growth direction. This kind of superstructure is of great importance during spinning, since most of the macromolecular chains must be oriented in the flow direction, and lamellae perpendicular to the flow contribute only slightly to the mechanical resistance of the filament.

2.4 Polymer behaviour during filament spinning

As molten polymers, or those in solution, are constrained to cross dies whose diameter is roughly a few hundred microns, it is necessary to adjust viscosities very precisely. The latter must be as low as possible in order to limit the pressures at the exit of the extrusion device. There exist several strategies to obtain a melt or a solution with adequate viscosity. For example, it is possible to decrease the molecular weight of the polymer, which appreciably increases the fluidity of the liquid material. However, this reduction in the molecular weight has important consequences for the final properties of the yarn, and in particular on its mechanical behaviour. The addition of a plasticizer makes it possible to decrease the thermal transitions of the polymer and increase its fluidity. Consequently, however, a significant deterioration in properties results, limiting the amount of plasticizer which may be incorporated. The processing temperature of the polymer is also a relevant parameter in decreasing its viscosity. Increase in temperature allows for the possibility of adjusting in a very precise way the viscosity of the mixture. It must not, however, exceed the thermal degradation limit of the material. It is easier to use this parameter to control viscosity with non-polar polymers (polypropylene, polyethylene) than with polar thermoplastics such as polyamide or polyester.

Determination of the extrusion and processing temperatures of a thermoplastic polymer is rather complex because the thermal effect is often regarded as not very important for bringing the polymer to the liquid state. The material is subjected to extremely high shear stresses, which strongly contribute to its plasticization and to its transition to the liquid state. Moreover, it should be noted that polymers generally have very low thermal conductivities. The low diffusivity of the material combined with its high viscosity tends to make the mechanical action dominate the thermal contribution. In single-screw extrusion, for instance, it is usual to consider the Brinkman number, Br, in order to define the relationship between the mechanical energy and the thermal contribution necessary to bring the polymer to a liquid state.

$$\mathrm{Br} = \frac{\text{Mechanical energy}}{\text{Thermal energy}} \qquad [2.2]$$

Formally, this ratio can be expressed by the following equation:

$$\mathrm{Br} = \frac{\eta V_F^2}{\lambda\left(T_F - \overline{T}\right)} \qquad [2.3]$$

where η represents the polymer viscosity (expressed in Pa.s), V_F is the screw linear velocity at the external top of the screw (in $\mathrm{ms^{-1}}$), λ is the thermal conductivity of the polymer (in $\mathrm{Wm^{-1}{}^\circ C^{-1}}$), T_F is the temperature of the sleeve (°C), and $\overline{T}$ is the average temperature of the polymer (°C).

If we consider an extruder with a screw of diameter 80 mm, running at a speed of 60 turns per minute, mixing a thermoplastic polymer of viscosity 10^3 Pa.s and thermal conductivity 0.2 $\mathrm{Wm^{-1}{}^\circ C^{-1}}$, and with a difference of 30°C between the temperature of the sleeve and that of polymer, we then obtain a value of 10.5 for the Brinkman number. This high value confirms that the major part of the energy provided to the system in bringing the material to the liquid state is of mechanical origin. The combination of the mechanical and thermal contributions then results in an increase in the temperature of the polymer during its processing. If this temperature is too high, the material evolves by thermal degradation. The generation of free radicals can also sometimes result in a disordered reticulation of the material, modifying its viscosity and the final properties of the yarn. The apparent viscosity can also become too low when the material crosses the dies, which disturbs the flow of molten polymer. If on the other hand the processing temperature is too low, extrusion becomes difficult, with flow irregularities and breakages of the filaments.

In order to limit the difficulties in finding an adequate compromise between the viscosity of the liquid polymer and preventing degradation, it is possible to accelerate the flow while applying intense pressures in the tools.

The pressures in the filtration systems and in the dies would then not be too high because the limits of mechanical resistance of the spinning system are likely to be reached. These pressures should, however, not be too low because they then give rise to a poor distribution of the polymer at the dies exit, and an irregularity in the quality of the multifilaments. In the industrial exploitation of a spinning process, requirements for regularity of the product are high (in particular in the case of continuous filaments), and several millions kilometres of yarns must be processed without breakage.

2.5 The die swell phenomenon

The optimization of the functional properties of a synthetic yarn (in particular from the point of view of its mechanical behaviour) requires maximum orientation of the macromolecules in the direction of flow. This objective can be difficult to achieve because of the generally pseudoplastic behaviour of such liquids. The dies which the polymers cross typically have the shape of a funnel (see Fig. 2.6). When the macromolecules are brought into liquid state in the system, they will undergo stresses because of the irregularity of the die's diameter. Under the action of the pressures at the exit of the extrusion device, the polymer will cross successively three zones in the die (see Fig. 2.6). The advance of the macromolecules from Zone I towards Zone II shows that the flow rate will increase because the section of the capillary

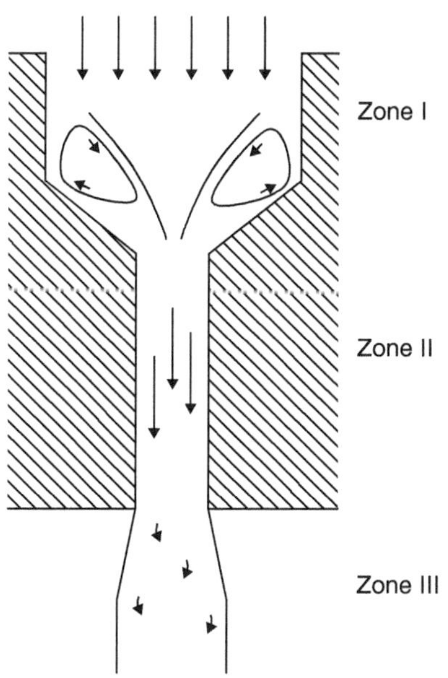

2.6 Typical evolution of the geometry in a capillary die.

2.7 Die swell phenomenon for a polylactide/polycaprolactone alloy.

decreases appreciably. The material flow being constant and the polymer being incompressible, this last is subjected to a side compression and an elongational stress. The macromolecules will thus tend to disentangle and become oriented to the flow direction. The pseudoplastic viscosity of the polymer will then decrease. When the liquid material reaches Zone III, i.e. the opening of the die hole, the macromolecules will be released to minimize their entropy. A strong tendency to return to the state of a ball is observed. A bulb is then formed below the die hole, known as the phenomenon of die swell. In these conditions, the extrudate has a diameter larger than that of the die. This very important phenomenon was first observed by Barusin 1893, but was more correctly analysed by Merrington in 1943 (Merrington, 1943). The Barus effect is defined by the following relation:

$$B = \frac{D}{D_0} \tag{2.4}$$

where D represents the diameter of the extrudate and D_0 the diameter of the die. This ratio of swelling B generally lies between 1 and 4, but can reach much higher values in the case of polymer alloys. As an example, Fig. 2.7 shows the aspect at the exit of a die (melt spinning) of a mixture of polylactide and polycaprolactone. The dies used in this case have a diameter of 400 microns, whereas the extrudate obtained shows a diameter of several millimetres. The ratio of swelling depends on several parameters, such as the residence time of the polymer in the stressed zone (Zone II), the length of the capillary, the molecular weight of the polymer, and the temperature in the die. Swelling can occur immediately after the exit of the die if the flow is relatively slow, or be shifted if the extrudate runs out at higher speed.

The measurement of swelling must be undertaken in a very delicate manner because of the stress relaxation after extrusion. Some time is necessary for the swelling to be maximized. Time (or the corresponding distance to the exit of the die) depends on the extrusion speed and on the relaxation time of the liquid polymer. Moreover, the tensions applied on the extrudate and its own weight are likely to modify the value of D. It is the same with regard to the temperature of the extrudate which, without special precautions, will undergo a strong cooling at the exit of the die.

The length L of the die in Zone II has also a very important influence on the swelling phenomenon at the exit of the die. Let us consider for example two extrusion experiments carried out with two dies of same diameters D_0, but different lengths L and $2L$. If the pressure exerted on each die is adjusted so that the same flow Q is obtained, we observe the production of rods with a diameter lower for the case of capillary length $2L$. The Barus ratio thus decreases as the length of the die increases. The reduction of B with L (or $\frac{L}{R}$) can be explained by the material relaxation in the channel of the die where this is sufficiently long. In practice, extreme cases are chosen by defining B_0 as the limit of the Barus ratio when $\frac{L}{R}$ tends towards zero, and B_∞ when it tends towards the infinite.

The flow rate in the capillary is another parameter which conditions swelling at the exit of the die. The Barus number in fact increases with the shear

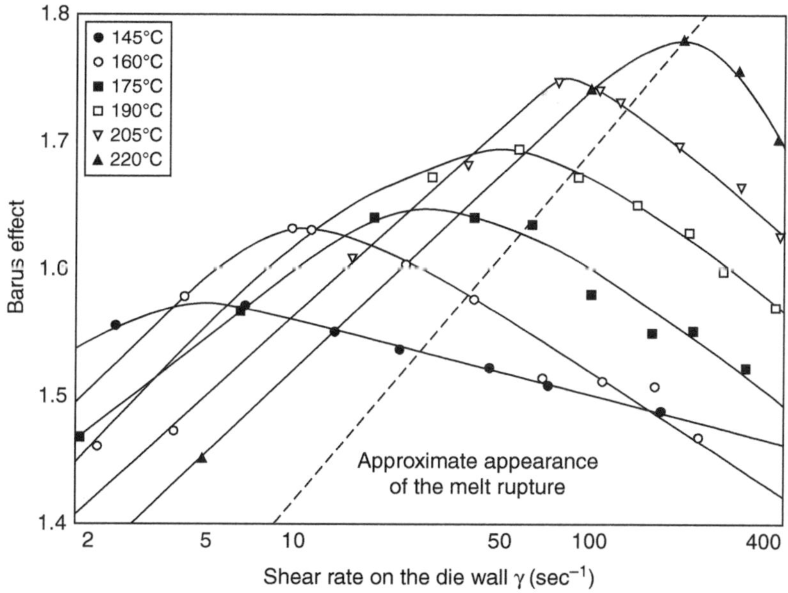

2.8 Evolution of the Barus number as a function of the shearing rate at the exit of a spinning die.

rate γ up to a breaking value, at which point certain flow defects appear (see Fig. 2.8). At this breaking value of γ, B passes through a maximum and then strongly decreases. This threshold, often called the 'rupture of melting' can be characterized by small irregularities on the extrude at surface (roughness, shark skin, etc.), by corrugated or helicoidal deformations, or by periodic deformations or breakage.

2.6 Drawing of polymers: changes in morphology and properties of filaments

At the die exit, and particularly in the case of melt spinning, the polymer is at a higher temperature than its melting point. Simultaneous with the swelling phenomenon at the die exit, the material is gradually brought back to room temperature. As the temperature decreases, the apparent viscosity of the polymer increases. Under the action of the tensile force, the polymer flow is refined and the filament diameter decreases until reaching its final value. This deformation occurs between the die and a point where the material reaches its glass transition temperature T_g, i.e. between a liquid and a rubbery state. Below this critical temperature, the polymer reaches a glassy state where it is comparable to a non-deformable material.

Between the molten state and the end of the rubbery state, the polymeric filament undergoes a very strong tension which compensates (at least partially) the Barus effect. At the exit of the die, the macromolecules relax and lose the orientation acquired in the capillaries. Under the action of the increasing tension, the molecules are aligned and oriented in the flow direction. As the temperature continues to decrease, it eventually becomes equal to T_g. The macromolecules cannot any more be relaxed, and the organization acquired previously is fixed. In certain zones of the filament, segments of macromolecules will be well aligned. In others, macromolecular entanglements persist. The state of organization will depend on the spinning conditions (cooling temperature profile, tensile speed, etc.).The yarns resulting from the spinning process thus do not show the characteristics of macromolecular orientation and crystallinity sufficient to obtain all the required properties. An operation of drawing is then required, which consists in applying a longitudinal deformation in a temperature range between the glass transition temperature of the polymer and a maximal temperature above which the orientation given to the yarn cannot be memorized any more. It is theoretically impossible to carry out fast deformations in a homogeneous way below T_g. The usual temperatures for the drawing of polyethylene terephthalate, for instance, lie between 80°C and 100°C, whereas they are between 50°C and 100°C for polyamide 6.6.

The drawing ratio generally lies between 2 and 4. This ratio can go up to 6 in the case of certain high performance fibres. Drawing is accompanied by a self-heating phenomenon, which can cause a temperature rise beyond the glass transition temperature without external heating of the yarn (in particular for polyamide 6.6). The phenomenon of striction also occurs which is generally fixed and located on a metallic part of the drawing device. The identification of the striction point should not be confused with the appearance of a constriction point obtained during the cold drawing of a static specimen. In a dynamic industrial drawing, the striction point is imposed by a mechanical device which constitutes the point of application of the drawing force where the temperature of the yarn reaches the glass transition temperature. The modulus decreases strongly and the striction occurs.

The macromolecular orientation and the crystallization which occur during drawing are still insufficient to lower the thermal contraction sufficiently. A supplement to crystallization is obtained through heat treatment, possibly accompanied by relaxation, which is carried out on the drawing device itself. This results in a disorientation of the amorphous phases, the macromolecules tending to return to the shapes of disorientated balls which are thermodynamically more stable. This complementary heat treatment is typically carried out between 130°C and 180°C for polyethyleneterephthalate, and between 150°C and 190°C for polyamide 6.6.

In recent decades, new generations of fibres have appeared, for instance, ultra-high molecular weight polyethylene (UHMWPE). The main interest in such materials lies is the fact that their mechanical properties are greatly increased. The major difficulty for the fibre processing of these polymers comes as a result of their very high viscosity in the molten state, which limits their processability by conventional methods. Due to this, important efforts have been undertaken using, for instance, gel-spun hot-drawing technologies. As macromolecular entanglements play an important role in the subsequent extension of the chains during drawing, it is necessary to dissolve the polymer in an appropriate solvent to favour their disentanglement. Drawing then promotes the slippage of the macromolecular chains, and drawing ratios of up to 60 or more may be reached (Pennings and Smook, 1984; Penning *et al.*, 1991).

2.7 References

Ajji A and Dumoulin M M (2006), 'Very high modulus high-density polyethylene and high-modulus polypropylene obtained by solid-state roll drawing', *J Appl Polym Sci*, **102**, 3391–3399.

Amornsakchai T and Songtipya P (2002), 'On the influence of molecular weight and crystallisation condition on the development of defect in highly drawn polyethylene', *Polymer*, **43**, 4231–4236.

Francesco A D and Duckett R A (2004), 'Effects of orientation on mechanical properties of uniaxially oriented polystyrene films', *Polymer*, **45**, 8005–8011.

Frank F and Tosi M (1961), 'On the theory of polymer crystallization', *Proc Roy Soc*, **263**, 323–339.

Fujimori A, Ninomiya N and Masuko T (2008), 'Structure and mechanical properties in drawn poly (L-lactide/clay) hybrid films', *Polym Adv Technol*, **19**, 1735–1744.

Gornick F and Hoffman J D (1966), 'Nucleation in polymers', *Ind Eng Chem*, **58**, 41–53.

Hosier I L, Bassett D C and Moneva I T (1995), 'On the morphology of polyethylene crystallized from a sheared melt', *Polymer*, **36**, 4197–4202.

Kalay G and Bevis M J (1997), 'The effect of shear controlled orientation in injection moulding on the mechanical properties of an aliphatic polyketone', *J Polym Sci Polym Phys*, **35**, 365–291.

Keller A and Kolnaar J W H (1997), 'Flow-induced orientation and structure formation', in *Materials Science and Technology, A Comprehensive Study*, 18, Weinheim: VCH, Chapter 4, 189–268.

Lagasse R R and Maxwell B (1976), 'Experimental study of kinetics of polymer crystallization during shear-flow', *Polym Eng Sci*, **16**, 189–199.

Lauritzen J I and Hoffman J D (1960), 'Theory of formation of polymer crystals with folded chains in dilute solution', *J Res Natl Bur Stand*, **64A**, 73–101.

Merrington A C (1943), 'Flow of viscoelastic materials in capillaries', *Nature*, 152, 663.

Murariu M, Doumbia A S, Bonnaud L, Dechief A L, Paint Y, Ferreira M, Campagne C, Devaux E and Dubois P (2011), 'High performance polylactide/ZnO nanocomposites designed for films and fibers with special end-use properties', *Biomacromolecules*, **12**, 1762–1771.

Nagasawa T, Matsumura T and Hoshino S (1973), 'Film forming process of crystalline polymer. 2. Microstructure', *Appl Polym Symp*, **20**, 295–313.

Penning J P, Dijkstra D J and Pennings A J (1991), 'Tensile force at break of gel-spun/hot-drawn ultrahigh molecular weight polyethylene fibres', *J Mater Sci*, **26**, 4721–4726.

Pennings A J and Smook J (1984), 'Mechanical properties of ultra-high molecular weight polyethylene fiber in relation to structural changes and chain scissoring upon spinning and hot-drawing' *J Mater Sci*, **19**, 3443–3450.

Peterlin A (1971), 'Molecular model of drawing polyethylene and polypropylene', *J Mater Sci*, **6**, 490–508.

Schoene A, Ziabicki A and Jarecki L (2005), 'Transient uniaxial orientation of flexible polymer chains in a wide range of elongation', *Polymer*, **46**, 3927–3935.

Solarski S, Ferreira M, Devaux E, Fontaine G, Bachelet P, Bourbigot S, Delobel R, Coszach P, Murariu M, Da Silva Ferreira A, Alexandre M, Degee P and Dubois P (2008), 'Designing polylactide/clay nanocomposites for textile applications: effect of processing conditions, spinning and characterization', *J Appl Polym Sci*, **109**, 841–851.

Somani R H, Hsiao B S, Nogales A, Srinivas S, Tsou A H, Sics I, Balta-Calleja F J and Ezquerra T A (2000), 'Structure development during shear flow-induced crystallization of i-PP: in situ small-angle x-ray scattering study', *Macromolecules*, **33**, 9385–9394.

Vargas A F, Orozco V H, Rault F, Giraud S, Devaux E and Lopez B L (2010), 'Influence of fiber-like nanofillers on the rheological, mechanical, thermal and fire properties of polypropylene: an application to multifilament yarn', *Compos Part A: Appl S*, **41**, 1797–1806.
Vleeshouwers S and Meijer H E H (1996), 'A rheological study of shear induced crystallization', *Rheol Acta*, **35**, 391–399.
Ziabicki A, Jarecki L and Schoene A (2004), 'Transient biaxial orientation of flexible polymer chains in a wide range of deformation conditions', *Polymer*, **45**, 5735–5742.

3

Technologies for the manufacture of synthetic polymer fibers

J. HAGEWOOD, J Hagewood LLC, USA

DOI: 10.1533/9780857099174.1.48

Abstract: From the first commercial beginnings of the fiber industry in the early twentieth century until today, the spinning of synthetic fibers has undergone remarkable changes. The early use of synthetic fibers in apparel has exploded into thousands of end products, from apparel to composite structures in aircraft to biodegradable sutures and beyond. Although the same basic principle of pushing a polymer melt or solution through a small hole to form a linear stream of material that is solidified and drawn into a fiber is the same today as it was in the beginning, the spinning speeds, polymers and types of processes have changed. The engineering and design of the equipment used to produce synthetic fibers is based on a delicate compromise between scientific understanding of polymers and fiber spinning processes, end-use requirements, and economics. Presently, there are two somewhat opposite trends driving fiber-producing equipment design and supply. One trend is focused on reducing the manufacturing cost of well-defined commodity fiber products such as polyester filament and staple. The other trend is focused on specialty fibers with unique properties that require new polymers and processes to meet the ever-changing demands of the marketplace. Examples of each of these trends are reviewed and discussed.

Key words: synthetic fibers, fiber extrusion, fiber melt spinning equipment, fiber solution spinning processes.

3.1 Introduction

Synthetic fibers are ubiquitous in modern day life. They literally surround us in our clothing, bedding, and home furnishings. They are also hidden in dust covers under sofas and beds, as filters in our heating and air conditioning systems, sound absorbers in our car doors, and stuffed in pillows and cushions. They are the primary ingredients in diapers for our children, wipes for cleaning, and many window treatments. They brush and floss our teeth and are used to stitch and bandage our wounds. Many surgical procedures would not be possible without the use of biodegradable synthetic fiber sutures that hydrolyze and dissolve in the body. Fibers are often used as reinforcement materials in plastic parts, automobile tires, and even concrete. In many ways, life in the twenty-first century is based on our understanding, production and use of synthetic fibers.

The engineering and design of the equipment used to produce synthetic fibers is based on a delicate compromise between a scientific understanding of polymers and spinning processes, end-use requirements, and economics. Most large fiber production plants are built by a relatively small number of engineering companies and equipment manufacturers based in Western Europe and Japan. As the installed capacity has moved to Asia, particularly China, these companies have also moved much of their engineering staff and production to Asia, to better compete with rapidly growing Asian competitors.

Presently, there are two somewhat opposite trends driving fiber-producing equipment design and supply. One trend is focused on reducing the manufacturing cost of well-defined commodity fiber products such as polyester filament and staple. This trend has resulted in ever larger spinning plants connected to continuous polymerization (CP) systems and ever faster spinning speeds. Today, a typical new fiber plant will have an annual capacity of over 500 000 metric tons (some are much larger), and can have filament winding speeds up to 8000 m per min for continuous filament yarns (Engelhart, 2011). The other trend is focused on specialty fibers with unique properties that require new polymers and processes to meet the ever-changing demands of the marketplace. This equipment is much smaller, slower and generally custom built. It is often built in-house by the fiber producer or end user. Obviously, it is impossible to describe all the equipment and processes used today to manufacture synthetic fibers.

In general, fiber spinning is divided into one of two processes: melt spinning where the polymer can be heated to a melt state without damage and extruded into fibers and quenched; and solvent spinning where the polymer is dissolved into a solvent to make a spinnable dope that is extruded into fibers, after which the solvent is removed. There are many variations on these two basic processes (in particular there are many types of solvent spinning processes). Examples of melt spun fibers include polyester, nylon, and polypropylene. Examples of solvent spun fibers include Kevlar®, Lycra®, rayon, and Lyocell®. Some polymers such as polyethylene are both melt spun and solvent spun, such as in the Tyvex® and Spectra® processes (Salem, 2001). What follows is only a sampling of the better-known commercial equipment and machinery used in this gigantic industry.

3.2 Textile filament and bulk continuous fiber (BCF) spinning

Filament spinning processes used to produce continuous fibers resulted in the largest percent age of synthetic fibers manufactured in 2010. These fibers were consumed primarily in textile and home furnishing applications, and amounted to 20.8 Mt (Engelhart, 2011). Although the largest portion of fibers spun by these techniques is polyester; nylons, polypropylene, and other polymers can also be spun this way. The larger lines are usually fed by a CP and may be part of a larger complex that supports the production of staple, continuous filament and polymer pellet production from the same polymerization unit. When the fiber is going to be used without the addition of bulking in some fashion, a single-step spin-draw operation is generally used. If the fiber is required to be bulked to meet certain end-use requirements, both single-step and two-step processes are available.

In the single-step unbulked process, the polymer is extruded through individual packs arranged in rows in a container referred to as a spin beam, which is often heated by a vapor phase liquid such as phenol/phenol. Each pack contains filtration media and a spinneret with a small number of holes. Typically the number of holes is between 30 and 100, although spinnerets with both smaller and larger numbers of holes are used. Each hole creates a filament, and the filaments from each pack are collected together into final threadlines as they pass through a quench zone cooled by air. If there are eight packs in a row, eight threadlines will be created, which will either be drawn by heated rolls or spun at very high speeds (greater than 5500 m per min) to produce a fiber with low elongation and high tenacity so that further drawing is not needed before use. The filaments are then consolidated by interlacing with high pressure air before being transferred to a winder with one threadline being wound per tube or package. Winders today can wind up to 16 or more packages at a time (Oerlikon Textile GMBH and CO., 2011). Recent innovations in this area have seen increases in the number of filaments per threadline, which have led to the production of smaller tex per filament fibers and higher winding speeds, which have eliminated the need for mechanical drawing of the fibers. Oerlikon Barmag has recently introduced a new high speed winder called WINGS, which also integrates fiber drawing and interlace with the winder (Fig. 3.1) (Susanne Beyer, February 2010). Other past innovations include larger package sizes, increases in number of packages per winder, and automatic doffing and handling equipment.

Often continuous filament fibers require some type of bulk or twist before they can be used in textile applications. This is accomplished in carpet yarns generally produced from nylon 6, nylon 66 or polypropylene by using a BCF machine (Oerlikon Textile GMBH and CO., 2011). These lines are similar to the one-step fiber lines described above, except that the spinning speed

3.1 Photograph of Barmag WINGS winder.

is much slower and the fiber tex is much higher. Most of these machines are also extruder fed. The filament shape is usually trilobal. Because of the lower spinning speed, the fibers are spun as unoriented filaments in the first step and are then drawn mechanically using heated godets. The fibers then pass to a stuffer box or bulker while they are being heated by steam. The bulked fibers are then transferred to a cooling drum where they are relaxed before being drawn off, transferred to a winder, and wound on packages under low tension. There have been few major technical changes in this operation in recent years. The process speeds have increased slightly and the number of ends per position and ends on a winder have also increased. Two major trends have been in the use of polyolefin polymers to make carpet yarns, and movement of the production of these fibers from traditional fiber producers to in-house production at carpet manufacturers such as Shaw and Mohawk.

Most bulked continuous filament fibers are produced for apparel applications and are processed using a two-step method. In the first step the fibers are spun as straight flat yarns at speeds between 3500 and 5500 m per min, which results in a fiber that is only partially oriented but fully crystallized. The fibers in this stage are extremely stable and can be stored for years without degrading; however, they are still not fully drawn. This final operation

3.2 Photograph of Barmag_eFK false twist texturing machine.

is later completed in a second step using a draw twist machine (Fig. 3.2). As named, this operation includes a drawing step, typically 1.6× for polyester, to complete the fiber orientation resulting in fibers with a final elongation to break in the range of 25–35%. Hundreds of thread lines are processed together on a single machine, which typically operates at a speed between 600 and 1200 m per min. Drawing takes place in a heated chamber between two sets of rollers operating at different speeds. After drawing, the yarn is passed through a twisting unit composed of multiple ceramic or plastic disks. The twist in the yarn builds up through the twist unit until a point at which the twist reverses itself and afterwards is heat-set in a second heated chamber. Because of the reversing twist produced in the yarn, this type of bulking is called false twist texturing. This process adds approximately 15% additional bulk in the yarn and helps to increase the cover in a fabric and create a softer and dryer hand.

3.3 Staple fiber spinning

The second largest volume of synthetic fibers produced annually in the world today is polyester staple fiber (PSF). The process and general design of the equipment used to produce these fibers have changed little since the early 1950s, when this process was first introduced to industry. What has changed dramatically is the size and location of the lines. PR China alone now accounts for 62% of all PSF production (Engelhart, 2011). The second largest producer is India; however, it accounts for only 3% of the worldwide production. Many new PSF lines being installed have capacities greater than 300 t per day as compared to the original lines, which had capacities more in the range of 10 t per day. This has resulted in a large decrease in the manufacturing conversion cost between the polymer ingredients and final baled cut staple. Another change is the consolidation and elimination of equipment suppliers, especially in the West. Presently, the only major Western supplier of large PSF lines is Oerlikon Neumag headquartered in Austria (Oerlikon Textile GMBH and CO., 2011). There are a number of suppliers in China, including Gangying Industry, JiangyinSelen Chemical Fibre Equipment, and Beijing Chonglee Machinery and Engineering, Company, to mention a few (Zhangjiagang Gangying Industry CO. LTD., 2011).

Polyester (or other fibers such as nylon or polypropylene) staple fibers can be produced by either a one-step process or a two-step process. The process steps are the same in both cases; however, the fiber spinning speed is lower in a one-step process and the fiber is drawn, crimped and heat-set in one continuous machine. These one-step lines generally have smaller capacities and are often fed by extruders. They are mostly used for specialty fibers, such as bicomponents and non-apparel end uses and nonwovens.

The larger lines all use a two-step process. A drawing of an older style small line of this type is shown below in Fig. 3.3 (Hills, Inc., 2011). The line shown is extruder fed from polymer chips and is bicomponent; however, most new lines of this type today are homopolymer polyester fed from a CP unit. The polymer is pumped, metered, and filtered from the CP unit to large round spinnerets that contain hundreds of thousands of holes in an annular ring pattern. The filaments are quenched from the inside out so that they are spread apart while being cooled. This is one of the newer equipment innovations that have led to the ability to build larger capacity lines. The fibers are oiled and then collected by turning rolls into tow, which is collected in containers. The filaments at this point generally have a dtex between 2 and 10 and will later be drawn as a tow from 2 to 6× to reach the final filament diameter. The fiber spinning speeds of these lines range between 600 and 1800 m per min. Again, the newer lines tend to run at the higher speeds versus the older lines, which often had cross flow quench and rectangular spinnerets.

The next piece of equipment in the two-step process is a staple draw machine. The tows from multiple containers are collected together and fed across cantilevered rolls into the machine where it is first drawn. Drawing takes place either between heated rolls or in a large water bath that is heated. In the larger lines these rolls may be over a meter in length. In the water draw systems used in the older lines, the drawing tended to be more even; however, the speed of water drawing is limited, which has led to the use of electrically heated rolls in the newer units. After drawing, the fibers are crimped by passing the tow band through a heated gear unit. The fibers are then collected on a moving open metal belt and heat-set in a large closed oven. For the final steps, the crimped fibers are cut and transferred by air to a baler unit, where they are pressed and wrapped into bales (DM and

3.3 Drawing of a Hills bicomponent staple spinning line.

E Corporation, 2011). Typical fiber cut lengths range from 3 to 6 cm. This length is easily adjusted by changing out the cutter head. The final bales are transferred by tow trucks or automatic handling devices to a warehouse to await shipment (Anon., 2011).

3.4 Industrial fiber spinning equipment

In the past few years the volume of fibers produced in North America and Europe for industrial end uses has exceeded that produced for other uses, such as apparel. Synthetic yarns for industrial end use have different requirements from those used for apparel and home furnishings. These fibers are continuous in length and generally have high tenacity, high modulus, and low shrinkage as compared to textile fibers for apparel. The other major difference is that these fibers are generally higher in both filament and yarn tex. As such, the spinning process, although similar to a single-step process for filament fibers, does have a number of differences. For large volume end uses such as tire cords, the processing line may be fed by a CP unit. For small markets, the lines are often extruder fed. The primary polymers used to make these fibers are polyester, nylon 6, nylon 66, and polypropylene. The molecular weight, and thus viscosity, of the polymers used for industrial fibers is generally higher (Hans A. Krassig, 1984) than for those used for apparel and other textile applications.

Another difference found in these lines is the size of the packs. The spinneret capillaries are larger. The quench units are also much longer, and often include delayed quenching. The spinning speed is traditionally lower, with a range from 1000 to 2000 m per min. These three differences are required to allow the fibers to cool and be extended over the longer residence time dictated by larger filament sizes and higher molecular weight polymers. After spinning and quenching, the fibers go through multiple stages of draw and relaxation while being heated above the glass temperature of the polymer. The most common process uses two stages of draw and one stage of relaxation at a high enough temperature to initiate crystallization.

The fibers are interlaced and generally wound on packages at speeds from 2500 to 3500 m per min. Because of the high tex of these yarns, doffing or removal of the packages from the winder occurs frequently. Little has changed in this process or the equipment used during the past few years (Nakajima, 2000).

3.5 Monofilament fiber equipment

Monofilament extrusion is a process used to make continuous strands of polymeric material for a variety of consumer and industrial products. Both one and two polymer systems called co-extrusion or bicomponent extrusion are widely used. Synthetic monofilament fibers are produced by extruding a melted polymer through small spinneret holes. This process is well known and is broadly applied in industry. The preparation, melting, and extrusion of the polymer are similar to those used for film blowing, die casting, and various fiber spinning processes. However, for higher tex fibers (diameters of >100 microns) the fiber processing is unique and is similar to slit film fiber processing (Hans A. Krassig, 1984). In the case of low tex monofilaments, the fiber processing steps are similar to the production of standard industrial multi-filament fibers.

Monofilament fibers are found in many applications. Some applications are easily seen and recognized; however, in most cases the casual observer will not be aware of the use of a monofilament fiber in a product. Three of the most obvious applications for monofilament fibers are fishing line, 'weed-eater' string, and toothbrush bristles. Fishing line is generally made from polypropylene because of its low density that allows the string to float in the water. 'Weed-eater' string and toothbrush bristles are generally made from nylon. Co-extruded products can be found in all of these applications.

Most thermoplastic polymers can and are used to make monofilament fibers. The most common material is high molecular weight polypropylene because of its low cost, ease of extrusion, and high linear strength when extruded and drawn into a fiber. The extrusion temperature for this material is generally in the range of 190°C, although higher temperatures can be used to lower the melt viscosity. Nylon 6, Nylon 66, Nylon 612, and various nylon copolymers are used in selective monofilament applications. Because of the relatively high cost of nylon polymers, they are used only where a required property, such as dyeability, high modulus, or high abrasion and fibrillation resistance is required. Nylons have higher melting points than polypropylene and are generally extruded at temperatures in the range of 260°C. The crystalline melting temperature of these polymers is around 230°C, and they have a glass transition temperature in the range of 70°C.

The process begins with melting of the polymer using a standard extruder. Single barrier flight screws with a Maddox mixer on the end are generally used, although twin screw extruders are also used. The melting process is mostly accomplished by shear heating from the turning of the screw. Multiple temperature-controlled zones are placed along the length of the extruder. In monofilament fiber spinning, the extrusion process is controlled by a pressure transducer at the exit of the extruder that is used to control the speed of the extruder screw to give fairly constant pressure of the melted polymer at the entrance to a metering pump.

Because high pressures and even fiber diameters are generally required for monofilament fiber spinning, a metering pump is placed between the extruder exit and the spin die. A typical pressure at the exit of the extruder is 7000 kPa (1000 psi) to 21 000 kPa (3000 psi), which is used to feed the inlet to the metering pump.

The polymer metering or gear pump is used to increase the pressure in the melt stream to the level needed to force the polymers at a constant volumetric rate through filtration media and the spinneret. Since the pressure across the filter media changes as the filter is used, the exit pressure from the melt pumps will vary; however, the flow rate must remain constant if the filaments are to have a consistent diameter. High pressure across the spinning holes is one of the keys to fiber properties and affects the molecular orientation and diameter of the filament. This process starts with high shearing of the polymer through the spinneret.

A standard gear pump can have one or more exit streams as needed. Each stream has a constant volumetric output based on the rotational speed (RPM) of the gears. The gear pump is driven by an AC motor with inverter or vector drive speed control. The exit pressure from a metering pump in a typical spinning process will range from approximately 7000 kPa (1000 psi) to 25 000 kPa (3500 psi). Once the pressure reaches 25 000 kPa (3500 psi), the process is typically shut down and the filter media replaced.

The last device in the extrusion process is the spin head. The spin head consists of a pump block, heating device, and spin packs. The metered polymer from the extruder flows to the metering pump and into the spin pack. A heated housing surrounds the pump block, metering pump, and spin pack to maintain all the equipment that comes into contact with the polymer at the desired temperature required for proper extrusion of the polymer.

The spin pack contains filter media such as screens or shattered metal, which is used to filter the polymer before it goes to the spinneret (Reifenhauser GMBH and CO., 2011). The spinneret contains one or more holes of precise diameter and length to spin the polymer and form a fiber with the right diameter. Since the filaments will be drawn in a further process, the diameter of the fibers at this point is not the same as the final diameter. For monofilament fibers, the extrusion speeds typically range from 50 to 500 m per min.

Once the filaments have been extruded, they must be cooled or quenched. For large diameter filaments (>100 microns) this is usually done in a water bath. When small diameter monofilament fibers are extruded, they can also be quenched in air chambers. The speed of the rolls in the water bath is a function of the fiber diameter, and the larger the diameter of the monofilament, the lower is the speed of extrusion. A typical spinning speed for large monofilaments is in the range of 30–100 m per min. If a large number of filaments are spun from a pack, there may be several rows of spin holes which are slightly staggered so that the filaments come together on the first roll in the water bath in a band or sheet of individual threadlines. From the first roll, the filaments travel to the second roll and exit the water bath. An air knife at the exit of the water bath is usually used to strip as much water as possible from the fibers before entering the draw stage.

The monofilaments after extrusion and quenching are still highly unoriented and need to be stretched in order to get the high strength (tenacity) required for most applications. The entire process is shown in Fig. 3.4. A comb guide is used to align the filaments through a series of rolls going at the same speed as the rolls in the water bath. The filaments then go through a hot air oven that heats the filaments to the point where they can be easily drawn or stretched. A second series of rolls is used at the exit of the oven to increase the speed of the filaments by 4–7 times. In most cases, two ovens and three sets of rolls are used for drawing. Each set of rolls is controlled by an inverter that controls the draw rates for different products.

After stretching, most polymers need to be heated to a temperature that will crystallize or heat-stabilize the filaments. This needs to be done under a different tension and at a higher temperature than used in the draw step. Therefore, after exiting the draw rolls the filaments travel through a final heat oven and exit into a final set of rolls. These rolls generally are run at a slightly lower (1–5%) speed than the draw rolls. This reduces the tension in the yarn as the heat setting occurs. The final properties of the filaments are now complete.

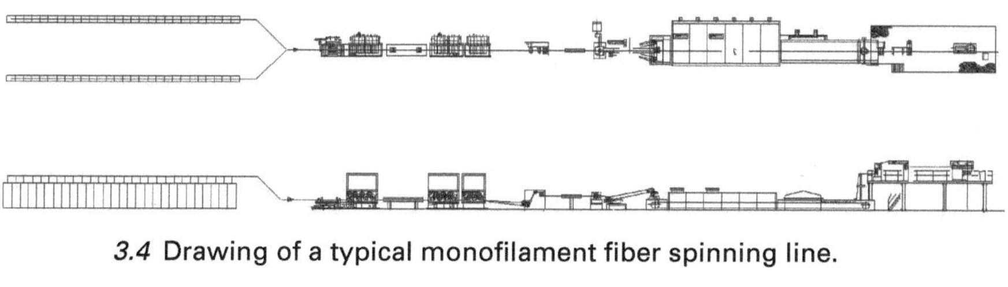

3.4 Drawing of a typical monofilament fiber spinning line.

The last step in the monofilament extrusion process is the winding of the filaments on individual tubes or spools. The winders can be arranged in a single bank or in multiple rows. The speed of the winders is individually controlled by a tension feedback device on each winder through which the filament passes. In most cases the winder spools are changed manually when the spools become full. Automatic transfer winders are available in the market, but due to the low processing speed of most monofilament lines, their cost may not be justified. After winding, the filaments are tested for properties and packaged for shipping to the customer.

3.6 Bicomponent fiber equipment

Bicomponent fibers have been a small but rapidly growing market for over 40 years. The growth rate has typically been over twice that of fibers in general and the sales price of the fibers has also been approximately twice the sales price of similar homopolymer fibers. All of the melt spinning processes described earlier can be modified to spin bicomponent fibers. All that is needed is two sources of polymer, such as two extruders, separate piping generally with individual heating controls, and bicomponent packs. The rest of the equipment and process, whether staple or filament, are similar. The largest manufacturers of bicomponent packs are Hills in the United States and Kasen in Japan (Kasen Nozzle Mfg. Co. Ltd, 2011). The internal components of the pack have been the major design element contributing to success in this area. The two polymers enter the pack in at least two different locations and are filtered separately before entering a series of channels that carry each polymer to the back entrance of the spinneret capillary, where they are combined into one unmixed polymer stream in each individual spinning hole before being extruded. There have been many significant developments in recent years in bicomponent pack designs. The major development has been the use of etched stacked plates to create the polymer channels in the packs. This has led to smaller packs with higher hole densities, larger number of segments and islands or splits in the fibers, and shorter residence times in the packs, which allows some temperature separation in the polymers during processing.

Bicomponent fibers can be divided into several different categories as shown in the chart in Fig. 3.5. These variants are discussed as follows (Hills, Inc., 2011).

3.6.1 Sheath and core products

A first type of sheath/core is self-bonding fibers consisting of a low melting temperature sheath and a higher melting temperature core. Common

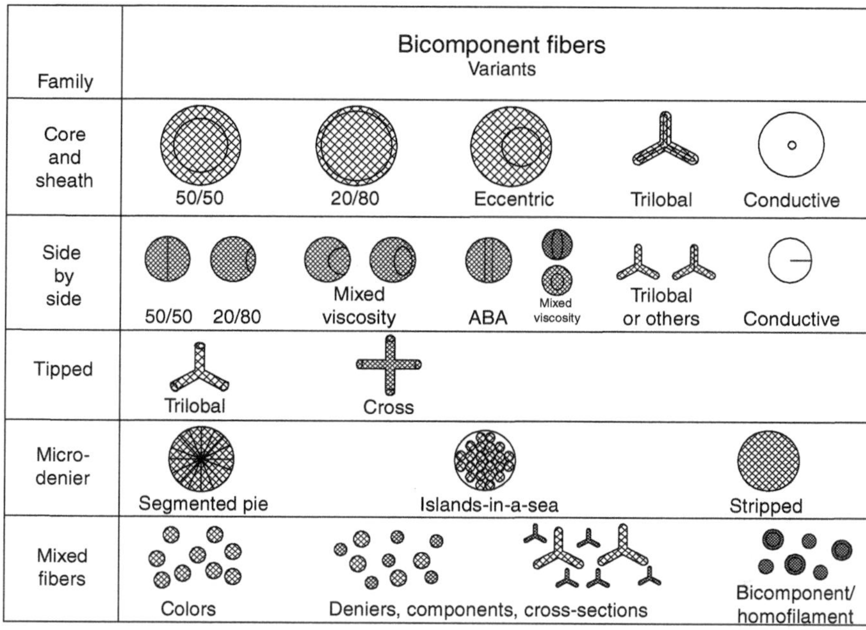

3.5 Hills chart of bicomponent fiber types. (*Source*: © HILLS, Inc., USA.)

sheath/core combinations in such applications include PE/PP, PE/PET, Co-PET/PET, and PP/PET. These products are used in 100% form, as well as in blends with homopolymer filaments. Staple fiber applications of these products in nonwovens are one of today's largest commercial uses of bicomponent fibers.

A second type of sheath/core is filled fibers consisting of a core polymer produced from recycled material, conductive material or other material that is covered by a sheath that possesses desired esthetics or other properties. A third type of sheath/core is a sheath polymer containing expensive additives or other attributes primarily useful on the surface of filaments.

3.6.2 Side-by-side products

Side-by-side products are usually used as self-bulking fibers. Self-bulking is created by two polymers within a filament having a different strain level or shrinkage propensity. Higher processing speeds, reduced equipment investment, more bulk, improved hand, enhanced aesthetics, and elimination of downstream processes are among the advantages seen with this technology. Side-by-side self-bulking fibers are now in fairly common use as stuffing in pillow cases, sleeping bags, and similar applications.

3.6.3 Tipped yarns

These fibers are used to produce special aesthetics, bonding, or other properties. One commercial application for this technology is in the production of capillary filters such as 'felt' tipped ink pens using a melt blowing type of process.

3.6.4 Microfiber products

Microfibers are produced by bicomponent techniques in which the fibers are either split apart, or one of the two components is dissolved or melted away. This technique has long been used to produce much smaller fibers than possible with homopolymer spinning, and is the technology behind such products as Ultrasuede® and similar fabrics (Kuraray CO. LTD, 2011).

3.6.5 Mixed fiber products

If a fiber producer owns equipment with two polymer capabilities, it is also economical to produce various types of mixed fiber products. The ability to control the level of mix from very light, to total, to random, is easily selected and managed. Examples of some commercial applications include the following:

- Color mixes to produce the desired aesthetics.
- Size and cross-section mixes to produce higher bulk or other desired aesthetics.
- Mixes of homopolymer filaments with bicomponents at varying levels to control certain fabric properties such as bonding strength. Some useful results can also be obtained by mixing of bicomponent cross-sections to obtain specific fiber or fabric properties.

It is very difficult to define the exact worldwide production capacity for bicomponent fibers. Most machines are extruder fed, and many machines in Asia were made locally and are not always successful in making good products. Also, many machines that have bicomponent capability are used to run standard homopolymer products. However, it is estimated that approximately 2–3% of worldwide synthetic fiber production equipment has bicomponent capacity. Of this capacity, approximately 23% for filament yarns, 31% for staple yarns, and 46% for spunbond. Approximately 75% of bicomponent fibers go to a nonwoven end use.

Staple bicomponent fiber capacity is fairly equally distributed throughout North American, Europe, and Asia. The lines are small as compared to standard staple lines. Two major products are produced. One is sheath/core fibers with a low melt sheath that is used for binder fibers in carded, thermally bonded nonwovens. The other product is side-by-side fibers (usually polyester and co-polyester) that are self-crimping and used for fiberfill. Production is approximately equal between these two products.

By far, the largest and fastest growing capacity for bicomponent fibers is in the area of spunbonds. There are approximately 50 such lines worldwide. About a third of these are pilot lines for product development, a third is older lines, and a third is modern curtain spinning lines with draw aspirators. It is estimated that 15–20% of all spunbond capacity worldwide has bicomponent capability. Much of this capability is not presently being used to make bicomponent products; however, intense activity to develop new bicomponent products is reported and is reflected in patent applications. The increase in bicomponent spunbond capacity is continuing at a rate of over 10% per year. This increase is fueled by both new lines and retrofitting of older lines. As new bicomponent products are developed, this activity is expected to increase rather than decrease or stay level. It is estimated that within a few years over half of all new spunbond machines installed will have bicomponent capabilities (Wilkie, August 2009).

3.7 Spunbond lines

Spunbond processes and fabrics have been commercial for over 40 years. The processes, fabrics, and markets are quite varied. The best known spunbond fabrics and markets are cover stock for diapers and feminine hygiene products. However, spunbond fabrics are also found in carpet backing, geotextiles, furniture dust covers, carriers for fabric softeners, and a host of other industrial and consumer applications. The key to most of these markets has been a low cost covering material with sufficient tensile and opacity properties to fit the application. The latest market trend is towards pre-wetted wipes for cleaning and similar uses. The product needs for this fabric include short term abrasion resistance and ability to hold solvents.

As the new century began, we saw new beginnings in spunbond processes and products. New techniques are being developed to improve uniformity and barrier properties of spunbond fabrics to allow spunbond to compete successfully with carded thermal bonded nonwovens. Fabrics from new polymers and polymer combinations are poised to enter the market. Tactile as well as tensile properties are being improved to the point where even apparel applications may be considered as a future market (Anon., 2011). Several of these trends are described in the following paragraphs.

3.7.1 Lower tex fibers

One approach to making spunbond fabrics with better cover and uniformity is to spin lower tex fibers. This has been an ongoing and gradual trend as spinning processes and polymers have improved with time. However, the recent use of high pressure draw jets has resulted in a distinct reduction in spunbond fiber tex without a resulting decrease in production. The key to this technology is high filament velocities. With the slot draw process, fiber dtex as low as 0.7 with polypropylene and 0.5 with polyester have been achieved. Fiber exit velocities are as high as 4500 m per min with polypropylene and 6000 m per min with polyester. High spinning speed allows the fibers to become fully oriented and crystallized, which results in excellent shrinkage properties as well as achieving low tex. This results in spunbond fabrics with good heat stability as well as good cover and uniformity. A schematic of a typical high pressure slot draw jet process is shown in Fig. 3.6.

One of the interesting features in the use of slot draw jets is the resulting compactness of the spunbond line. The main difficulty in obtaining high fiber speeds is air drag. This is overcome with the draw jet concept by using a short quench chamber and by locating the draw jet directly below the quench chamber. This reduces the distance between the spinbeam and the belt to one to two meters. The downside of this technology is that cross-machine fiber properties and fabric uniformity must be built into the machine from the beginning as there are only a few process parameters to adjust that affect

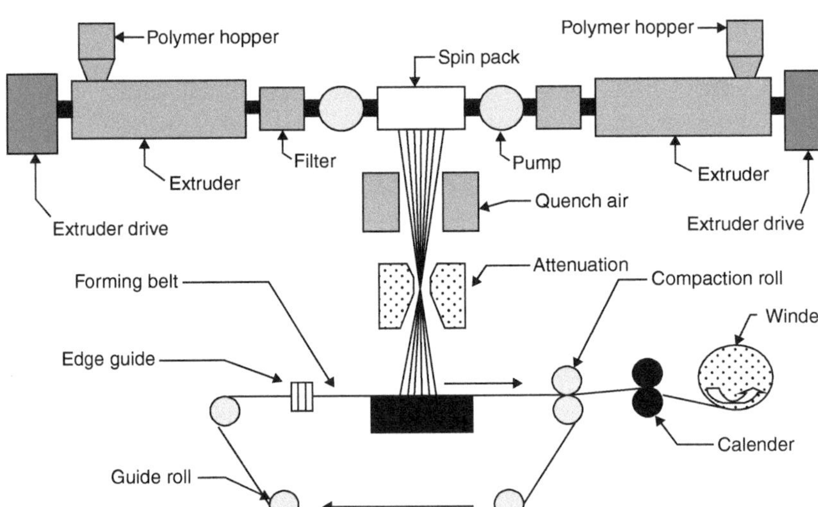

3.6 Drawing of the bicomponent spunbond process.

fiber laydown, as compared to the older gun draw process in which the exit of the fibers from the gun had multiple location adjustment possibilities.

3.7.2 Bicomponent fibers

A rapidly growing development in spunbond processes is the use of bicomponent fiber technology. This trend is based on Hills, Inc. bicomponent spinning technology, which has been used for staple and filament spinning for many years (Wilkie, August 2009). Commercial implementation has been ongoing for over 10 years. This includes both new lines and retrofitting of older lines. This area continues to be very active with new lines that include at least one bicomponent beam for a spunbond nonwoven line being ordered worldwide approximately every 2 months.

There are several advantages in spinning sheath/core bicomponent fibers for spunbond fabrics. The first is that use of a lower melt temperature polymer on the sheath allows the bonder to run at a lower temperature and a higher speed. This increases the line speed and reduces manufacturing cost. The second advantage is that the bonding temperature does not affect the orientation of the core polymer. This results in higher fiber orientation and therefore higher fabric strength. A third advantage is that special effects such as softness and hydrophilicity can be imparted to the fabric from the sheath polymer without affecting other fabric properties that would be affected if the fabric were formed from fibers made totally with the sheath polymers. A further advantage is that high cost additives can be mixed in with only the sheath polymer, thereby greatly reducing the cost that would be incurred if the additive were put in a homopolymer fiber.

Other types of bicomponent fibers that split, such as segmented pies, are also being developed for spunbond fabrics. Once the splitting occurs in the fiber or in the fabric, the fiber dtex can be as low as 0.1 in spunbond and the tactile property of the fabric approaches that of conventional woven and knitted goods made from low dtex fibers. Much of this work is aimed at the apparel market, polishing cloth, and artificial leathers. The latest technology in this area is the use of a dissolvable polymer such as Exceval® that washes out with hot water leaving only tiny fibrils from the dissolved filament (Kuraray CO. LTD, 2011).

3.7.3 New polymers

The activity level to develop new polymers for fiber spinning is at an all-time high. This seems incompatible with the present low selling price of polyester and polypropylene, but many major chemical companies continue to look for polymers for special niche markets. Among these polymers are many varieties of co-polyesters, co-nylons, melt spinnable elastomerics, PLA, Polytrimethylene terephthalate(PTT), and others. When spun into a spunbond fabric these polymers, either in a bicomponent sheath or homopolymer fiber, can bring new and exciting properties to the spunbond world. Some of the most interesting of these new products are stretchable fabrics.

3.7.4 Hydroentangling

Hydroentangling, which can be described as needling a nonwoven fabric with very high pressure water jets, has been used as a process to produce fabrics since the 1960s. The process is also known as spunlace because of the physical appearance of some of the fabrics produced. The process is highly versatile and is particularly used by PGI to produce a line of lightweight open structure fabrics under the trade name Miratec® and by DuPont to produce barrier fabrics for medical uses under the trade name of Sontara®. Generally, the process starts with staple fibers that allow a variety of both natural and synthetic fibers to be used. Recently, there has been increased activity to apply hydroentangling directly to spunbond fabrics rather than using a carded staple fiber product. One such development in this area that uses bicomponent fibers resulted in the commercial product Evalon® by Freudenberg.

3.7.5 Open vs closed systems

When using a slot draw jet in a spunbond line there are two basic approaches to accelerate the fibers during spinning. One approach is to use high pressure in the drawjet to create a supersonic air flow in the throat of the ventura of the draw jet. This type of process uses a large quantity of air, and the area between the draw jet and the laydown table is open, which allows excess air to escape into the room. Filament velocities up to 10 000 m per min have been reported when the draw jet is moved close to the spinneret to reduce the air drag on the fibers. The negative side of this process is the difficulty in balancing air flow in the spinning area and maintaining good uniformity in the fabric.

The other approach is to use a closed slot draw system. The best example of this is in Reifenhauser lines. In this approach the system is essentially sealed or closed from the spinneret to the laydown table. Flow through the slot comes from the quench above and the low pressure created by the vacuum box below the belt. This system generally results in fabrics with excellent uniformity once the air flows are balanced correctly. However, the filament velocities that can be obtained are much less than with the open system.

3.7.6 Reicofil 4

A new Reicofil 4 spunbond line was introduced at Index 2002 in Geneva (Fi-Tech, Inc., 2011). The Reicofil 4 machine is a closed system; however, the obtainable velocities were increased to approach those of a closed system (Reifenhauser GMBH and CO., 2011). The hole density in the spinneret was also increased and bicomponent capabilities were added. The results of the higher hole density and higher filament speeds was a 50% increase in throughput per meter of beam length. Alternately, the fabric producer can keep the throughput the same and go to lower filament tex to produce softer and more uniform fabrics.

3.8 Melt blown processes

The melt blown process (Fig. 3.7) is a one-step process that converts polymer resin into low diameter fiber nonwoven web or tow (Andreas Desch, February 2011). The melt blown process, and its variants, is the only large-scale commercial process that is presently being used to directly produce melt spun fibers with diameters in the submicron range without splitting or chemically dissolving away polymer. It is a nonwovens operation that directly creates fabrics of various widths and thicknesses on large rolls. The products are mainly used in filtration applications, barrier fabrics, oil absorption mats, and battery separators. It is a rapidly growing business segment due to a large increase in the demand for better filtration and membrane medium. The major polymers used in the process are low molecular weight olefins.

The key to the melt blowing process is the spin head. The basic principle is the extrusion of low viscosity polymer melt through a single row of very fine holes placed close together in the order of 1000–4000 holes per m. These

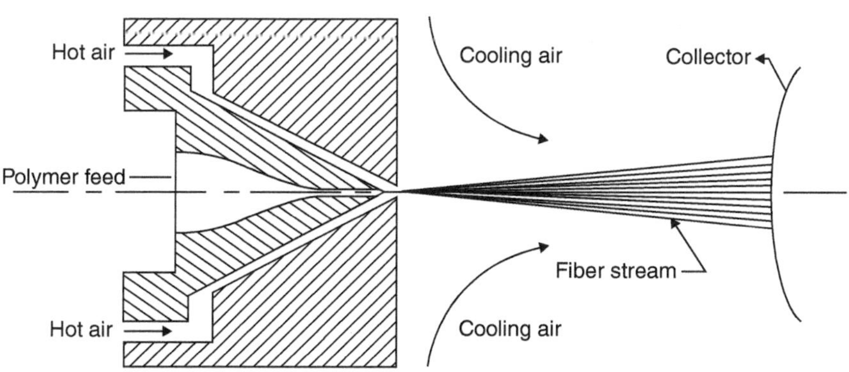

3.7 Schematic of the melt blowing process.

holes are usually drilled or made by fusing together two plates that contain etched channels. High velocity hot air is blown at the exit of the holes from both sides of the row of holes at an angle. This air keeps the molten polymer hot and attenuates the molten polymer into a fine fiber. At the same time a flutter is developed in the air stream that causes the fibers to flap rapidly back and forth as they are cooled by ambient air becoming entrained in the process. Finally the fibers are collected on a screen belt or drum.

The fibers produced by this method are generally very weak with low tenacity and modulus. One reason is that low molecular weight (low viscosity) polymer is required to make the process work well. The other factor is that the hot air keeps the polymer in the melt state as it is being attenuated, which is necessary because of the rapid acceleration of the polymer as it exits the spin hole. This results in low axial molecular orientation in the fiber as it is being formed.

There are several issues that influence the melt blown process which are different from other major fiber spinning processes. The first is the design of the extruder screw which is commonly manufactured specifically for the low densities and low viscosities associated with the polymers used in the process. The polymer must also be filtered to a greater level than required in most other fiber spinning processes. Hot compressed air controlled to an exacting temperature must be provided to the process and the angle of insertion and volume (air gap) of the air must be controlled precisely over the length of the spin head. The use of a single row of holes greatly restricts the fiber mass output of the process per meter of web width. This results in production and cost issues. Several attempts with limited results have been made to design heads that solve this issue and still produce low diameter filaments.

3.9 Solvent spinning lines

Solvent spinning is typically used when the polymer being spun into fibers has a decomposition point near or below the melting point of the material. There are also cases such as polyethylene where a solvent is used in a melt spinnable polymer to spin high molecular weight polymers such as Spectra® or to get other desirable properties such as in Tyvek®. The first commercially successful synthetic fiber was rayon which was spun from cellulosic based dopes. Today, even though apparel synthetic fibers are mostly melt spun, there are many successful solvent spun fibers in production. They are generally used in higher end applications where their unique properties justify a higher price. Some of the more common of these fibers are Kevlar®, Lycra®, Nomex®, acrylics for carbon fiber processing, Lyocell®, and cellulose acetates. The overall processes used to spin each of these fibers is as various as the fibers themselves.

There are two common elements in spinning solvent spun fibers. One is the mixing of the polymer with a solvent in a reactor or large mixer. The solvent component can be as high as 95% of the dope and as low as 5%. Generally, the dope is heated and fed to the spinneret using a metering pump identical to those used in melt spinning. Sometimes the dope is fed directly to the metering pump by pressure, but occasionally the dope is fed from an extruder to the metering pump. The other common element in all solvent fiber spinning is the extraction of the solvent. This begins to take place immediately after the fibers leave the spinneret. In a process called dry spinning this extraction is accomplished in a heated chamber where the solvent is flashed off and removed (an example of this is cellulose acetate). In a process called wet spinning, the spun dope in the form of a fiber is pulled through a series of water baths or other solvents. Drawing of the fibers is usually done before all of the solvent has been completely removed. These types of operations have been applied to many different processes including nonwovens, filament spinning and staple production.

Although solvent spinning allows us to manufacture fibers with properties that cannot be obtained with commercial melt spinning processes, there are several drawbacks to using this process. One such issue is that the removal of a solvent from a dope spun into a fiber takes time. This results in a much lower speed of fiber spinning. Typically, the spin speed in a solvent spun operation is an order of magnitude lower than that achieved in melt spinning. The equipment is also much more complex. Both of these factors directly impact the cost of capital and labor per unit of fiber manufactured resulting in a higher manufacturing cost. The second issue is the solvent itself. Because of the volume and cost of the solvent which must be used, but which does not become a part of the fiber, a sophisticated solvent recovery system is required. Solvents as a rule are also not environmentally friendly. Any chemical that dissolves polymers is going to be an issue when it comes to its effect on machinery and people. As a result, with the exception of acrylic fibers used for carbon fiber production, most cellulosic and acrylic based fiber spinning is no longer practiced in the US.

3.10 Hollow fiber membranes

Hollow fiber membranes are generally solvent spun and as such can be considered a sub-group of solvent spinning. Major polymers used in these processes are polysulfone, Polyvinylidene diflouride(PVDF) and polyethersulfone. The major applications are hemodialysis, water purification, and air separation. This is a huge business that has been growing at over 10% per annum for many years. Although the product usage in annual tons is very small, the end-use revenue is in billions (Anon., 2011).

A hollow fiber membrane is a fiber with an external diameter between 200 microns and 2 mm. The fiber is hollow in the middle with a continuous and tightly controlled open area. Unlike tubing, the side wall of the fiber is porous and generally contains large pores near the center open area that gradually become smaller as you reach the outside of the fiber. In some cases this pore structure is reversed. In either case there is a thin separation layer from 25 to 100 nm thick on either the outside or inside of the fiber against the open core that is a semi-permeable membrane through which diffuse takes place in the final product usage. It is the size and distribution of these pores in the separation layer that are critical to the use of the membrane. The pores characteristics are controlled primarily by the chemistry of the dope and the processing conditions used to manufacture the fiber. As an example, in hemodialysis blood is passed through the middle of the membrane and the pores in the separation layer allow only blood particles with a size below the diameter of the pores to pass through the membrane wall. By using a fluid medium on the outside of the membrane that contains additives that match the concentrations of blood particles smaller than the pore diameter, these particles can also be blocked resulting in an extremely selective system that mostly separates only waste products from the blood during dialysis.

Membrane spinning begins with dope preparation. Unlike standard solvent spinning, the dope usually contains several ingredients including polymer, solvent, and other additives. In a generalized process, the polymer concentration is between 20% and 30 % and the dope viscosity is between 1000 and 10 000 poise at room temperature. The dope is heated to somewhere between 50 and 100°C and fed from a pressurized supply tank to a standard metering pump or pumps. Most processing lines spin multiple threadlines with an individually metered dope steam to each spinneret hole. The processing speeds are between 5 and 75 m per min and there is no draw in the process once the fiber has completed its initial draw down between the spinneret and the coagulation tank. The spin hole contains a small tube through which a lumen fluid consisting of water, solvent, or a mixture of the two is metered. From the spinneret the membrane drops through a short air gap and into a coagulation tank where the solvent and dope rapidly separate. The fiber is then passed through successive rinsing tanks, dried, and wound on a spoke winder. The winder reel has a distance between the spokes slightly longer than the fiber length needed to fill a module. After the required number of fibers has been collected, the fibers are switched to another reel, and the first reel is removed. The fibers are then cut between the spokes. The resulting bundles of fibers are placed in a circular module and the ends potted. The potted ends are later cut across to expose the open cores of the fibers and end caps are placed on the module, which is now ready for use.

3.11 Gel spinning

Recent development of a type of solution spinning called gel spinning has resulted in commercialization of several high performance polyethylene fibers most notably Spectra® and Dyneema®. These fibers have tenacities approaching that of Kevlar® (3 GPa) but with lower modulus and much lower melting temperature. The process uses an ultra-high molar linear polyethylene (UHMWPE) that is dissolved in solvent. The ratio of solvent to polymer is generally in the 1–5% range. This dope is metered from the mixing unit through a multi-hole spinneret and into an extrusion tank usually filled with water. As the dope is cooled and solvent is extracted, gel-like filaments are formed in a type of network arrangement. These as-spun filaments, still containing large amounts of solvents, are then transferred to an oven and very slowly drawn up to 50× or more. The resulting fibers are then wound on packages (Salem, 2001). The equipment and machinery used for manufacturing these products has to date been built in-house by the fiber producers and is a closely guarded secret.

3.12 Nano-fiber spinning

There has been a lot of recent interest in developing research and commercial processes and equipment to directly spin fibers with diameters below 100 nm. Most of this work has involved a process called electro-spinning, and has taken place in various universities around the world. Although some researchers have reported success using melt spun polymers, most of this work has involved a form of solution spinning where the polymer is first dissolved in a solvent with a fairly high ratio of solvent to polymer, generally 90–70%. The major characteristic of this type of spinning is that a glass tube containing a small opening at one end is used as the spinneret and a high electric field (30–50 kV) is created between the dope in the spinneret and the collector. The collector is usually a roll of previously formed nonwoven fabric (either melt blown or spunbond) that is slowly moved across a metal plate used to create the electric field. The tiny fibers formed are then collected on the fabric and can be used as a high flux/high capture filtration medium.

The fibers exit the spinneret in a spiral motion as they move through space toward the collector, which is somewhat analogous to what we see in melt blowing. The diameters of the fibers generally vary along their length. The fibers are also very weak since the drawing operation is accomplished under low stress and occurs while the fiber is highly solvated, and so little molecular orientation is achieved. However, the biggest issues with this process are the extremely low production rates and solvent collection and recovery. Despite these issues, at least two large filtration companies have developed commercial operations using electro-spinning in the production of high end filters for air purification and personnel protection devices.

To overcome the low production rates possible with a single spinneret, Elmarco in Europe has commercialized a line of electro-spinning units where the dope is placed in a pan containing a rotating cylinder with multiple spikes. The small filaments are extracted via the electric field from dope collected on the spikes and they collect on a fabric slowly moved across a metal plate located above the dope pan. (Elmarco, 2009). There have also been numerous attempts to develop similar processes to produce nanofibers with high production rates. Most of these processes involve modifications of the melt blowing process or combination of electro-spinning and melt blowing.

3.13 Conclusion

Synthetic fibers, beginning with rayon in apparel, have expanded into multiple fiber types and products found throughout our lives. The processes and techniques used to produce these fibers have also changed dramatically during the past 70 years. Fiber manufacturing equipment and processes continue to change to both reduce the cost of fibers and to introduce newer fibers with specialized properties.

3.14 References

Andreas Desch, F. R. (February 2011). Dienes and rieter OFT Develop new nanofiber spinning installations. *International Fiber Journal*, **25**(1), 38–39.

Anon. (2011). Nonwovens Industry. (Online) Available at: www.nonwovens-industry.com.

Anon. (2011). Marketing Technology Service, Inc. (Online) Available at: www.marketingtechnologyservice.com.

Anon. (2011). Textile World. (Online) Available at: www.textileworld.com.

Anon. (2011). Filtration + separation. (Online) Available at: www.filtsep.com.

Anon., (2011). Filtration news. (Online) Available at: www.filtnews.com.

Beijing Chonglee Machinery Engineering Co., Ltd. (2011). Staple and filament spinning lines. (Online) Available at: www.ctamp.com.cn.

DM and E Corporation. (2011). Cutters for staple fiber lines. (Online) Available at: www.dmcutter.com.

Elmarco (2009). Nanospider equipment. s.l.:s.n.

Engelhart, A. W. (2011). *The Fiber Year Gmbh,* Speicher, Switzerland: Oerlikon Textile.

Fi-Tech, Inc. (2011). North American representative of for equipment manufacturers. (Online) Available at: www.fi-tech.com.

Hans A. Krassig, J. L. H. F. M. (1984). *Fiber Technology.* New York: Marcel Dekker, INC.

Hills, Inc. (2011). Bi-component spinning lines of all types. (Online) Available at: www.hillsinc.net.

Jutter Stehr, M. R. (February 2011). Barmag 'e-save' label offer energy savings and more. *International Fiber Journal*, **25**(1), 34–36.

Kasen Nozzle Mfg. CO. LTD. (2011). Bi-component pack manufacturer. (Online) Available at: www.kasen.co.jp.

Kuraray CO. LTD. (2011). Microfiber fabric producer. (Online) Available at: http://www.kuraray.co.jp.

Nakajima, P. T. (2000). *Advanced Fiber Spinning Technology*. Cambridge, England: Woodhead Publishing Limited.

Oerlikon Textile GMBH and CO. (2011). Filament spinning line and draw twist texturing units. (Online) Available at: http://www.barmag.oerlikontextile.com.

Oerlikon Textile GMBH and CO. (2011). Staple fiber and BCF Lines. (Online) Available at: http://www.neumag.oerlikontextile.com.

Reifenhauser GMBH and CO. (2011). Major producer of spunbond and melt blowing equipment. (Online) Available at: http://www.reifenhauser.com.

Salem, D. R. (2001). *Structure Formation in Polymeric Fiber*. Munich: Hanser Publishers.

Susanne Beyer, O. B. (February 2010). WINGS – The Oerlikon Barmag concept that revolutionized POY spinning plants. *International Fiber Journal*, **25**(1)20–21.

Wilkie, A. E. (August 2009). Advanced nonwovens equipment and technology from hills. *International Fiber Journal*, **24**(4), 20–23.

Zhangjiagang Gangying Industry Co. Ltd. (2011). Staple and filament spinning lines. (Online) Available at: www.gangying.com.

Part II

Spinning techniques

4
Melt spinning of synthetic polymeric filaments

A. RAWAL and S. MUKHOPADHYAY,
Indian Institute of Technology Delhi, India

DOI: 10.1533/9780857099174.2.75

Abstract: Melt spinning is among the most widely used methods for producing polymeric filaments. A range of polymers, namely poly(ethylene terephthalate), polyurethanes, polyolefines and polyamides, are generally melt spun. This chapter discusses the fundamentals of melt spinning, modeling of polymeric flows inside the screw and the spinneret, interaction of melt-spun filaments with the environment, and the limitations of melt spinning. Some recent developments and future trends pertaining to melt spinning are also discussed.

Key words: extruder, filament, polymeric fluid, melt spinning, spinneret.

4.1 Introduction

Melt spinning is one of the most popular methods for manufacturing polymeric filaments. While there are several methods available for filament production, melt spinning is the most economical approach due to the absence of solvents and the simplicity of the process. In melt spinning, the polymer pellets or granules are fed into an extruder consisting of a screw for melting by means of heat, and then the polymer melt is pumped through a spinneret under pressure. The extruded polymer is then quenched with cold air and the molten mass is solidified into filaments. The spun filaments lack adequate strength for industrial applications. Hence melt spinning is generally followed by the mechanical drawing of the extruded filament resulting in alignment of molecular orientations along the filament axis. This results in improved physical and mechanical characteristics. The mechanical drawing of filaments consists of many fold elongations of the filaments (starting from 2×), which can be achieved directly after spinning or carried out separately with undrawn extruded polymer as an input material. Polymers such as poly(ethylene terephthalate), polyurethanes, polyolefines and polyamides are generally melt spun. For a basic review of melt spinning, the reader is referred to Ziabicki (1976), Fourné (1999), and Beyreuther and Brunig (2007). We discuss some of the important principles, limitations, and recent developments involved in the formation of filaments through melt spinning.

4.2 Fundamentals of melt spinning

A typical melt spinning process requires a constant mass flow rate of molten polymer, which is maintained by a metering or a spinning pump generally positioned inside the spinning head, as shown in Fig. 4.1. The molten polymer is channeled into a number of individual capillary holes or slots of defined shapes and sizes present inside the spinneret. Each orifice is responsible for individual filaments. Subsequently, the long continuous filaments extruded through the spinneret orifices are cooled-off, solidified, and collected on a winder. The take-up speed is kept higher than the average

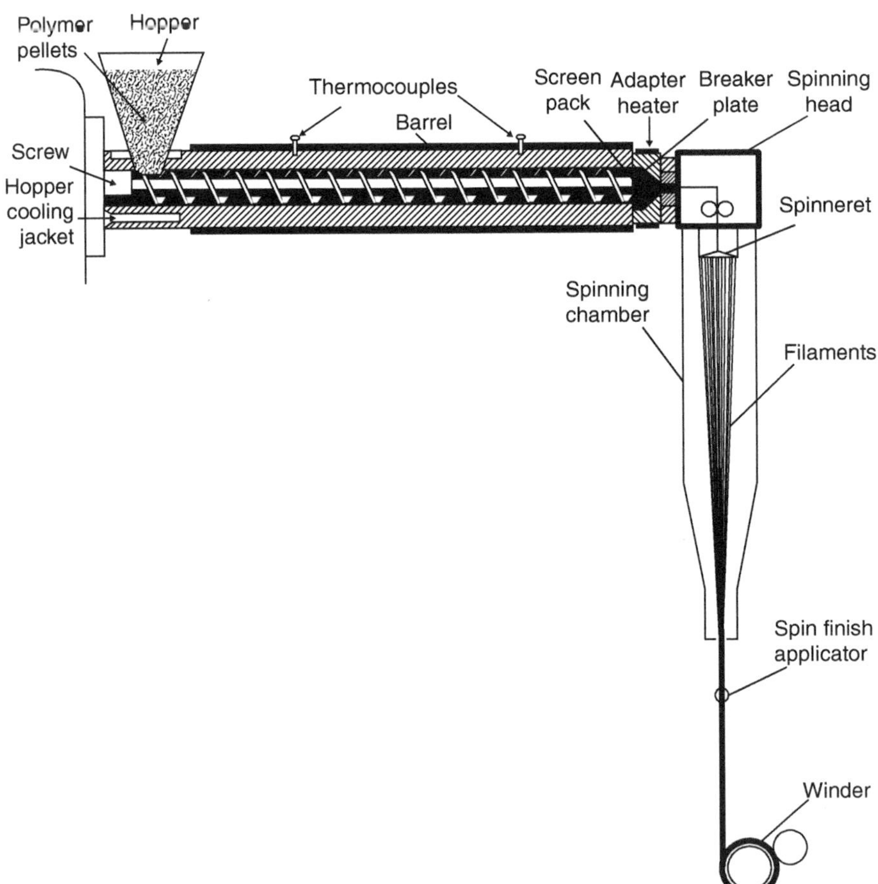

4.1 A typical melt spinning process.

extrusion speed or throughput rate at the spinneret, and the ratio of take-up speed (v_f) to average extrusion speed (v_0) is defined by the draw down ratio (DDR). Hence,

$$\text{DDR} = \frac{v_f}{v_0} \qquad [4.1]$$

In order to improve their structure and mechanical properties, the drawing of the filaments can further be carried out in a single- or two-stage process between take-up of the as-spun fiber and the final winding roller. Ideally, the filament formation in terms of *spinnability* requires the following conditions (Beyreuther and Brunig, 2007).

- Attenuation and acceleration of the filaments (DDR > 1)
- Production of continuous filaments with no filament breakage
- Steady-state conditions, i.e. filaments with uniform properties along the length

In general, the spinnability of the polymeric melt is determined by the maximum possible DDR for a given polymer. As indicated above, steady-state conditions are a prerequisite for filament formation through the melt spinning route. These steady-state conditions are fundamentally dependent on the physical balance equations of mass, energy, and momentum as discussed below (Ziabicki, 1967; Beyreuther and Brunig, 2007).

4.2.1 Mass balance

The mass balance equation states that there is no mass exchange taking place between the filament and the environment. Mass conservation is expressed in Equation [4.2] where Q is the mass throughput rate, A the filament cross-sectional area, ρ the polymeric density, and v the axial filament velocity:

$$Q = Av\rho = \text{constant} \qquad [4.2]$$

In the field of textiles, the product of filament cross-sectional area and its density is defined in terms of fineness (T_t). The diameter of the filament (D) can be related to its fineness as shown below.

$$D = 11.3\sqrt{\frac{T_t}{\rho}} \quad \text{for } T_t \text{ in dtex} \qquad [4.3]$$

or $$D = 9.44\sqrt{\frac{T_t}{\rho}} \quad \text{for } T_t \text{ in denier} \qquad [4.4]$$

where D is the filament diameter (μm) and ρ the polymer density (g/cm^3).

Ziabicki (1967) has characterized the process variables by means of a continuity equation in terms of *primary*, *secondary*, and *resulting* variables. Primary variables are the parameters that affect filament structure and its properties in addition to the process conditions of the melt spinning process. Secondary variables are obtained from the primary variables using the continuity Equation [4.2]. Resulting variables are determined by primary variables and the dynamics of filament formation process.

4.2.2 Energy balance

The energy balance equation represents the changes in temperature profile of the filament from the point of exiting the capillary holes to the points of solidification and take-up. Assuming a simple case of heat convection, the heat balance of any given volume element between x and $x + dx$ is given by the temperature gradient of the filament as shown below (Ziabicki, 1967; Beyreuther and Brunig, 2007).

$$\frac{\mathrm{d}T}{\mathrm{d}x} = -(T - T_{\mathrm{air}})\,Nu\,\frac{\pi\lambda_{\mathrm{air}}}{Qc_p} \qquad [4.5]$$

where T_{air} is the temperature of the surrounding air, c_p the specific heat capacity of the polymer, Nu the Nusselt number, and λ_{air} the heat conductivity of air.

4.2.3 Momentum balance

The momentum balance equation reveals the forces acting on the filament formed in a melt spinning process. The magnitudes of these forces play an important role in determining the structural development, specifically the structural arrangement of the polymeric chains. The force balance for a defined cross-section of the spinning line at distance x from the spinneret is shown below (Ziabicki, 1967).

$$F_{\mathrm{ten}}(x) = F_{\mathrm{Rheo}}(x) + F_{\mathrm{in}}(x) + F_{\mathrm{surf}}(x) + F_{\mathrm{aero}}(x) - F_{\mathrm{grav}}(x) \qquad [4.6]$$

where F_{ten} is the external tensile force, F_{grav}, F_{in}, F_{surf}, F_{aero}, and F_{Rheo} are the gravitational, inertial, surface tension, skin-friction, and rheological forces, respectively.

Details of the computation of the above forces are given in Ziabicki (1967) and Beyreuther and Brunig (2007).

The above equations have formed the basis for simulation of filament formation and have been extensively used and modified by many research groups (Beyreuther and Brunig, 2007). Furthermore, in order to understand the melt spinning process and the structural and the mechanical behavior of extruded filaments, it is imperative to understand models of polymer flows inside the screw and the spinneret.

4.3 Modeling of polymer flows in melt spinning

In general, polymer flow can be classified into two categories: tangential and axial (Rawal, 2002; Worth, 1979). Tangential flow is generated either by rotational motion of the cylinders relative to one another or as a result of the applied pressure gradient. Axial flow occurs when one of the cylinders moves longitudinally relative to the other, but can also be generated as a result of the pressure gradient. These flows can be combined together at different polymer processing stages.

4.3.1 Modeling of polymer flow inside the screw

In a melt spinning process, the two different types of polymer flow start to develop as soon as the polymer pellets or granules come into contact with the screw. Initially, tangential flow is developed due to the rotation of the screw but axial flow may also occur due to the pressure difference across the duct, or when the polymer flow comes into contact with the screw helix (Tadmor and Gogos, 1979).

Various theoretical models have been proposed to study polymer flow inside the screw for purposes of screw design. The first flow theory was given by Rowell and Finlayson (1922, 1928), in which the screw flights were assumed to be stationary and the barrel had the same velocity as that of the screw tip. Experimental results have shown that these assumptions are incorrect (Choo, *et al.*, 1980). Therefore, Campbell *et al.* (1992) designed a special screw extruder to study the drag flow. Here, the screw extruder, the barrel, screw core, and screw flights were rotated separately or in pairs. It was observed that the flow generated by the rotation of the screw was different from that obtained by rotating the screw flights. Hence, in real extruders, where the screw is rotating and the barrel is stationary, these boundary conditions need to be considered. Li and Hsieh (1996) have formed models for both drag and pressure flow by considering these boundary conditions.

The new analytical models yielded encouraging results; however, the models were valid only for Newtonian fluids, and as such cannot be used for real polymeric fluids (which are primarily non-Newtonian).

An alternative numerical approach was used to compute the flow of polymer inside the screw extruder (Middleman, 1965). The technique is based on the assumption that the polymer flow is a combination of drag and pressure flow in a straight rectangular duct. The assumptions behind this model are based on the well-known parallel plate analogy (Tadmor and Gogos, 1979) where the relative motions of the barrel and the screw flights are considered parallel. There are two motions for the flow of polymer when it passes through the screw, a circular motion which occurs as a result of the rotation of the screw, and a helical motion produced by the combined effect of the stationary barrel and the moving screw flight.

Booy (1963) has considered the effects of curvature and helicity of the screw and derived a model known as the *curved channel model*. This model was further generalized by Yu and Hu (1998) to calculate the velocity profile and flow rates. The results were compared with the parallel plate models and found to be very similar in drag and pressure flows. However, these models were more accurate in the case of deeper screw channels such as twin-screw extruders or some special single-screw extruders. The details of modeling of polymer flows in various stages of the screw are given by Tadmor and Gogos (1979).

4.3.2 Modeling of polymer flow inside the spinneret

After passing through the screw, the molten polymer flows into the capillary holes or slots of the spinneret, where it is pushed into a small duct to achieve the desired shape and size as determined by the spinneret. In general, the hole or slot shapes may be classified in two categories: circular and non-circular. In the case of circular shapes (Bird, 1977) the equations of motion involving non-Newtonian flows can be solved by assuming polymeric flows as power law fluids. The flow rates of such fluids are shown in Equation [4.7].

$$Q = \frac{\pi R^3}{1/n + 3}\left[\frac{(P_O - P_L)R}{2mL}\right]^{1/n} \qquad [4.7]$$

where Q is the volumetric flow rate, R the radius of slot, P_O the entrance pressure, P_L the exit pressure, L the depth of slot, and m and n are power law constants.

For non-circular shapes, the equations of motion may result in non-linear partial differential equations, which are difficult to solve analytically. Therefore, approximate methods such as the variational method (Kantorovich and Krylov, 1958) are generally used for solving non-Newtonian flow problems. Schechter (1961) used the application of the variational method to solve the non-linear partial differential equations of pressure drop and flow rate of the polymer for non-circular shapes such as a rectangle or square. Moreover, Mitsuishi and Aoyagi (1969 1973) used similar methods for other non-circular shapes such as an isosceles triangle. The results were based on the Sutterby model (1966), which incorporates a viscosity function based on the rheological constants. Flow curves with pressure drop and flow rate for both circular and non-circular shapes were generated and the results were compared with the power law model.

Arai and Toyoda (1970) formed a mathematical model to compute the wall shear stresses in a rectangular spinneret shape using a power law:

$$\frac{AB}{A+B}\frac{\Delta P}{A+B}=\frac{B}{A+B}\frac{2\int_0^{B/2}\tau_w(y)\mathrm{d}y}{B}+\frac{A}{A+B}\frac{2\int_0^{A/2}\tau_w(x)\mathrm{d}x}{A} \qquad [4.8]$$

where A, B, L are width, height, and length of the rectangular slot respectively, ΔP is the pressure loss within a length L due to the viscosity resistance of the flowing liquid, $\tau_w(x)$ and $\tau_w(y)$ are the shear stresses close to the wall surfaces at $y=\pm B/2$ and $x=\pm A/2$ respectively.

When a molten polymer flows into the small channel of the spinneret slot from a larger channel, a convergence is formed by the contraction of the channel cross-section, and relative elongation along with shear deformation are thus produced. Because of the small length of the slot, stresses due to deformation are not completely relaxed before the melt leaves the slot. This results in extrudate swell as demonstrated by measuring the cross-sectional area of extrudate, which is greater than that of the respective slot size through which it is extruded, a phenomenon known as die swell (see Fig. 4.2). The die swell ratio is the ratio of the area of the extrudate to the area of the die, and normally it is 1–2.5, although in extreme cases it can be as high as 8 (Ziabicki and Kedzierska, 1960).

The first die swell theory was presented by Tanner (1970) to compute the die swell ratio for given material properties. The die swell ratio was calculated for a very long die (length-to-diameter ratio approaching infinity) and was formulated using the normal stress difference and shear stress at the wall.

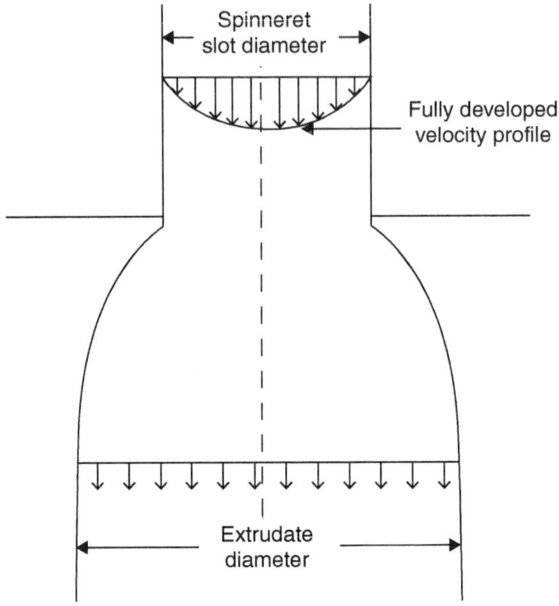

4.2 Polymeric fluid flows emerging from the spinneret slot illustrating the die swell effect.

$$\frac{D}{d}=\left\{2\int_0^1\left[1+\left(\frac{N_1}{2\tau}\right)^2\right]\xi\mathrm{d}\xi\right\}^{1/6} \qquad [4.9]$$

where $\xi \equiv \frac{2r'}{d}$

N_1 is the normal stress difference, τ the shear stress at the wall, r' the radial distance, d the spinneret diameter, and D the extrudate diameter.

Subsequent work took into account the velocity and shear field adjustments at the end of a die (Nickell *et al.*, 1974). White and Roman (1976) developed a modified formulation of Tanner's theory where the melt recovers from Poiseuille flow in a tube into a state of uniaxial tension. These theories have been validated by Vlachopoulos *et al.* (1972). However, the die swell theory was also studied using macroscopic mass and momentum balances given by Metzner and others (Metzner *et al.*, 1961; Graessley *et al.*, 1970), where the balance on the fluid between the capillary exit and flow downstream formed a relationship with the die swell. The first normal stress difference is shown in Equation [4.10].

$$(P_{11}-P_{22})_{R,L}=\frac{\rho D_0^2}{64n}\left(\frac{8V}{D_0}\right)^2\left\{(n+1)\frac{3n+1}{2n+1}-\left(\frac{D_o}{D_e}\right)^2\left[(n+1)+\frac{\mathrm{d}\log D_o/D_e}{\mathrm{d}\log 8V/D_o}\right]\right\} \quad [4.10]$$

where $(P_{11}-P_{22})_{R,L}$ is the primary normal stress difference, D_o the tube diameter, D_e the extrudate diameter, V the average velocity of the fluid, n the power law constant, and ρ the melt density.

The above equation is valid only in cases where the Reynolds number (which signifies the extent of turbulence in the fluid flow) is high. Therefore, a modified Metzner theory was proposed by incorporating an assumption that the pressure at the center-line is non-zero, which is valid for polymeric melts (Mori and Funatsu, 1973).

Die swell has also been related to the process parameters and flow conditions. For a given rate of shear, the die swell decreases as the length-to-diameter ratio of the flow channel increases (Arai and Aoyam, 1963; Bagley *et al.*, 1963; Buchdahl *et al.*, 1951). Die swell also decreases rapidly with an increase of length-to-depth ratio (L/D) of the slot as the polymer molecules are relaxed in a longer slot, with the result that the die swell value is lower than expected (Mori and Funatsu, 1973). It has been observed that as L/D approaches 20, the die swell becomes constant. This can be attributed to full development of the polymer flow. Moreover, the die swell is higher in square-shaped dies than circular dies (Mayer *et al.*, 1997). While in square-shaped slots the swelling takes place equally in both directions (i.e. depth and width), the swelling of filament produced from circular slots takes place only around the circumference. Therefore, on attenuation under a defined take-up speed, the filament area reduces from both directions equally, resulting in a minimum filament cross-sectional area (Rawal, 2002; Rawal and Davies, 2005; Rawal and Davies, 2006). Die swell can also increase with an increase in the screw speed for a single-screw extruder (Rejon and Alvarez, 1987). Furthermore, the die swell increases with the molecular weight, especially for linear polymers, which are less compact than long chain branched molecules (Racin and Bogue, 1979).

Die swell has also been investigated using models that are widely used for rubber-like solids. The analogy between polymer melt elasticity and rubber elasticity is based on the fact that both have long molecular chains (Treloar, 1958).

The most widely used equation to calculate die swell is shown in Equation [4.11] (Tanner, 1970).

$$\chi=\left(1+\frac{1}{2}S_R^2\right)^{1/6} \quad [4.11]$$

where χ is the die swell and S_R the recoverable shear strain.

The above die swell model is based on the assumption that the polymer fluid behaves similarly to a rubber-like solid. However, it is difficult to measure recoverable shear strain under general conditions.

Wong and Liang (1997) formed a relationship between melt flow index and die swell ratio using a power law. However, this approach requires information related to the shear stress and shear rates of polymers.

Numerical techniques such as finite element and finite difference methods have also been used for predicting the die swell (Duda and Vrentas, 1972; Richardson, 1970; Tanner *et al.*, 1975; Chang *et al.*, 1979).

Chin and Shyu (1988) have predicted die swell of polymers based on the assumption that polymeric fluids obey Criminate–Ericksen–Filbty (CEF) equation. They obtained an equation for die swell that depends on the processing conditions and thus helps to choose such for the required die swell. Similarly, Mitsoulis (1999) used the streamline finite element method to compute the die swell using different slot cross-sections such as circular, square, and cross-shaped sections. The fluid models used were Newtonian, viscoplastic (Bingham), and second order fluid. In addition to the finite element methods, the finite volume method has also produced very good results (Xue *et al.*, 1995).

Die swell is an important parameter for the design of slots used in polymer processing operations. For non-circular slot geometries the extrudate changes not only in dimension, but in shape as well. Predicting the die swell accurately is of great importance in designing the size and shape of the slot. However, the complexity of viscoelastic fluid models and the difficulties imposed by boundary conditions further intensifies the problem of die swell prediction, even for the simplest slot geometries. Therefore, the problem of die swell is still under investigation and is an active area of research, especially for slot geometries with non-circular cross-section.

In the past, surface tension was observed as a major factor responsible for the filament cross-sectional shape (Ziabicki, 1967; Takarada *et al.*, 2001; Rawal and Davies, 2005; Rawal and Davies, 2006). However, an increase in the cooling rate along with a higher viscosity of the melt can significantly reduce the deviation of the filament shapes from their corresponding slot geometries.

4.4 Filament interaction with the environment in melt spinning

As mentioned in Section 4.2, the energy balance equation dictates the development of the filament temperature from the point of emergence from the capillary holes to the points of solidification and take-up. This transition of molten polymer to the solidification of the filament is essentially related to heat transfer from the surface of the filament to the ambient medium, i.e. air in a typical melt spinning process. Hence, the non-dimensional constant, Nusselt number, (defined as the ratio of convective to conductive heat transfer coefficients) has been introduced in the energy balance equation. Generally, three heat transfer mechanisms from the filament surface to the surrounding medium can occur: radiation, free convection, and forced convection. In a melt spinning process, the contributions from radiation and free convection are negligible (Ziabicki, 1967; Beyreuther and Brunig, 2007). Filament cooling in a melt spinning process is primarily related to forced convective heat transfer. The heat transfer equations for forced convection are based on the empirical data found for stationary filaments in a moving medium rather than for axially moving filaments in a stationary medium (Ziabicki, 1967). Nevertheless, Kase and Matsuo (1965) have formulated an empirical relationship for the Nusselt number for forced convection (Nu_f) including the effect of cross-flow of the cooling medium, as shown in Equation [4.12].

$$Nu_f = a\left(\mathrm{Re}_{\mathrm{Parr}}^2 + b\,\mathrm{Re}_{\mathrm{Perp}}^2\right)^c \qquad [4.12]$$

where $a=0.42$, $b=64$, $c=0.167$, and $\mathrm{Re}_{\mathrm{Parr}}$ and $\mathrm{Re}_{\mathrm{Perp}}$ are Reynolds numbers related to parallel and cross air flow, respectively.

Using Equation [4.5], the Nusselt number can be recalculated by accurately determining the filament temperature along with the filament diameter and/or its velocity assuming that crystallization has not occurred. Most publications do not describe the effect of quenching cross air flow on the air friction coefficient. This may be because the air flow perpendicular to the filament axis changes the axial symmetric boundary layer surrounding the filament and leads to an additional momentum flux (Beyreuther and Brunig, 2007). More recently, it was demonstrated that in a high-speed melt spinning process, the axial air flow is more important than the cross air flow, while maintaining the same freeze-point stress and residual drawability (Sumesh *et al.*, 2010). It was also observed that replacing cross by axial air flow can increase productivity by 32%. However, in the case of cross air flow, a strong quenching air profile can cause a high cooling rate, which can lead to a short filament formation zone. On the other hand, increasing the quenching air flow results in a higher air friction coefficient (Beyreuther and Brunig, 2007). This indicates that the total air friction force and the resulting filament stress at the solidification point can increase or decrease with an increase in quenching air flow, and the resulting properties of the filaments under each condition may differ.

The filament interaction with the environment in multifilament spinning is not same as that in monofilament spinning. In multifilament spinning, the boundary conditions vary with the filament location within the filament bundle. Initially, there is tendency for hot filaments to dissipate heat to the quenching air thus increasing the overall temperature of the surrounding environment. Moreover, the transverse component of air velocity is significantly reduced as the air flow tends to follow the axial direction of the running filaments. The cross-flow of air also has a tendency to disturb the axial air flow, and the boundary layers of the individual filaments may overlap, resulting in the vanishing of friction forces for the individual filaments at the converging point. Thus, the multifilament spinning process can be treated as a two-phase system in which the filaments are embedded in the environmental matrix (Ziabicki *et al.*, 1998). The general conditions of heat and momentum, however, need to be satisfied at the boundary layers of these two phases.

Recently, a complex two-phase model of melt spinning has been proposed, which has accounted for effects such as crystallization and necking as well as with the environmental conditions, namely, air drag, inertia, gravity, and surface tension (Dhadwal, 2011). There were a number of challenges in such a complex model, ranging from selecting an appropriate initial guess value to solving highly difficult equations.

The multifilament melt spinning process is not limited to filament production; its principle has formed the basis of a spunbonding nonwoven process. The major difference between the filament and spunbonded nonwoven production is that in the latter process the filament take-up is not realized using godets or a winder but is achieved instead by means of air friction, as shown in Fig. 4.3 (Jeon, 2001). Here, the take-up speed due to air drag is not known from the beginning, and the higher the velocity of the axial air stream the higher the velocity of the filament due to the higher friction and drag forces transferred to it. Hence, at constant throughput rate, higher filament velocity reduces the filament diameter. The filament velocity, diameter and other related properties are dependent on the component of air drag transferred to the filament and the contribution of other forces, including rheological, inertial, and gravitational.

The principles of the melt spinning process have been explored for other nonwoven processes, such as the melt blown process. Here, the polymer is extruded through a spinneret consisting of several slots, and streams of hot air rapidly attenuate the extruded polymeric fluid into fine fibers as shown in Fig. 4.4 (Bhat and Malkan, 2007). Subsequently, the attenuated fibers are blown by high velocity air onto a collector conveyor, self-bonding the fine fibers. Recently, theories pertaining to the formation of polymer jets in a melt blown process have been proposed revealing that bending

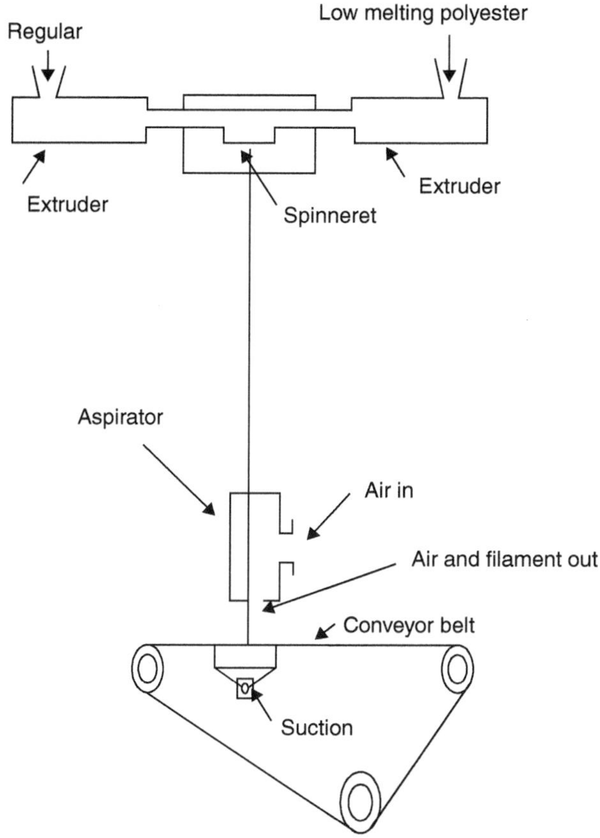

4.3 Schematic diagram of spunbonding process (Jeon, 2001).

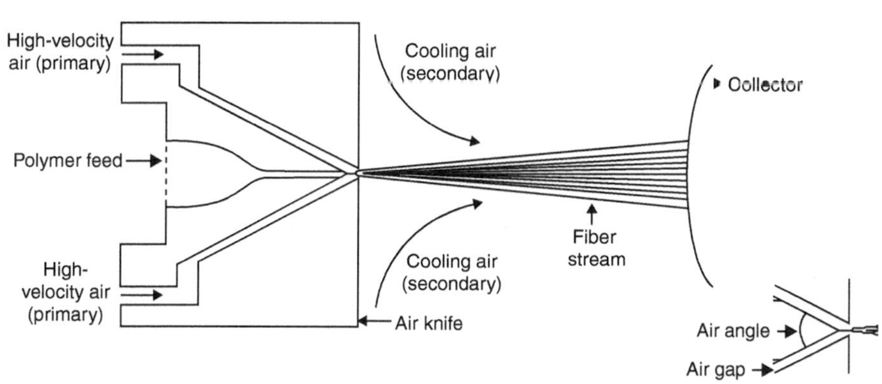

4.4 Web formation in a melt blown process (Bhat and Malkan, 2007).

perturbations of the polymer melt jets are triggered by the presence of surrounding turbulent eddies, and enhanced by the drag force acting on the jets (Yarin *et al.*, 2010). These theories have also been extended for melt blown consisting of multiple jets (Yarin *et al.*, 2011). However, simulation of polymer flows inside the slots of the spinneret of the melt blown is yet to be achieved.

4.5 Limitations of melt spinning

The limits of the melt spinning process need to be considered for developing filaments with extremely low or high diameters. For coarser filaments, the highest possible mass throughput rate per hole of the spinneret is required while maintaining the lowest possible take-up velocity. Higher mass throughput rate per hole requires a longer distance to the solidification point. Moreover, the filament needs to be cooled down below the solidification temperature before it comes into contact with the godet or winder; otherwise fibers may stick to the godet or each other. Polymeric fluid extruded at a very high throughput rate can exhibit rough or highly irregular surfaces and can even break up into individual fragments, a phenomenon known as *melt fracture* (Ziabicki, 1967). In some cases, air quenching has been replaced by a water quenching method, which has been found to be a useful technique for producing significantly coarser filaments.

Filaments with low diameters after drawing can be produced by keeping the minimum possible throughput rate, maximum possible take-up velocity and a high draw ratio. Increasing the take-up velocity increases the filament orientation, whereas a decrease in the mass throughput rate decreases the filament diameter. The increase in the filament orientation is not significant. However, higher take-up velocity results in a higher spin line stress and, as polymeric fluids are viscoelastic in nature, this can lead to filament breakages due to brittle cohesive fracture. It has been observed that a reduction of throughput rate is more effective than increasing the take-up velocity for obtaining finer filaments (Beyreuther and Brunig, 2007). Hence, the physical limits of the melt spinning process are dictated by an increase in take-up velocity and/or a decrease in the throughput rate.

4.6 Developments in melt spinning

Melt spinning is not limited to the production of conventional polymeric filaments; various developments in terms of type of filaments, structure–property relationships, introduction of nano materials, and advances in process characterization have evolved in the last few years. Some of these developments are briefly discussed below.

4.6.1 Structure development in composite filaments

Structure developments in melt spinning have been researched extensively, although most work has been carried out on homopolymers (Lu and Spruiell, 1987; Jinan C, *et al*, 1989; Choi and White, 2000). Rwei (2000) experimented with blends of polyethylene terephthalate (PET) and polyethylene naphthalate (PEN). The polymers were melt spun with a high-speed winding process in a single-screw extruder combined with a spinning setup. The filaments had a single melting point and glass transition temperature, which indicated excellent compatibility in both the amorphous and crystalline phases. Birefringence and wide-angle X-ray measurements indicated that compounding PEN into PET suppressed stress-induced orientation and decreased stress-induced crystallization in the filaments. It was observed that the addition of PEN to PET relaxes the formation of skin-core structures for as-spun filaments and reduces the occurrence of broken filaments. Interestingly, although the addition of PEN reduced levels of crystallinity, it did not affect the tenacity and shrinkage of the compounded filaments. The elongation of the filaments could be reduced by 30–40%, eliminating the need for further drawing. PEN has a rigid backbone, and the results might be due to decreased crystallinity. Adding PEN to PET improves PETs spinnability during high-speed spinning.

Kim and Kim (2006) used poly(ethylene 2,6-naphthalate) (PEN) and poly(ethylene terephthalate) (PET) as composite filaments reinforced with a thermotropic liquid crystal polymer (TLCP) and used a melt blending and spinning process to synthesize high performance filaments with improved processability. The researchers observed an increase in the birefringence and density of the TLCP/PEN/PET composite filaments with the spinning speed and attributed this observation to enhancement of the molecular orientation and effective packing between chains in the composite filaments. The annealing process resulted in the formation of a more ordered and perfect crystalline structure in addition to higher crystallinity. These composite filaments had enhanced mechanical properties. The increase in the crystallite size and the degree of chain extension with increasing spinning speed resulted in a gradual increment of the long period for the composite filaments.

4.6.2 Incorporation of nanotubes in melt spinning

Hooshmand *et al.* (2011) blended polypropylene/polyamide with carbon nanotubes (CNTs) and melt spun to as-spun and drawn filaments. Thermal analysis showed that increasing the polyamide content decreased the degree of crystallinity in the blends. Characterization of the filaments demonstrated that both conductivity and tensile strength were improved by increasing the amount of polyamide in the blends as well as the melt blending temperature. This work involved the preparation of compounded PP/COPA 9 (low melting point co-polymer of polyamide) blends containing 5% CNTs and subsequent melt spinning into monofilaments. Thermal analysis proved that the CNTs increased both the crystallization temperature and the thermal stability of the blends as compared to neat materials. In addition, use of a higher temperature in the melt blending process and the addition of 1% compatibilizer enhanced the conductivity and tensile strength of the filaments. The authors ascribed this observation to the formation of a more homogenous structure. The same improvement was observed by increasing the amount of COPA in blends. The researchers achieved the highest conductivity with the sample consisting 30% PP–65% COPA–5% CNTs with 1 phr (parts/100 rubber) compatibilizer prepared at 240°C. It was not possible to increase the amount of COPA to higher than 65% in the composite. A comparison between as-spun filaments and drawn filaments showed that although the mechanical properties were improved by drawing, the electrical conductivity was decreased by several orders of magnitude. The authors showed that by drawing the filaments, the distance between the nanotubes increased and the percolated network was destroyed. The maximum conductivity and lowest percolation threshold for composite filaments occur when the fillers orient randomly, resulting in closer carbon nanotubes in the filament structure.

Pötschke *et al.* (2010) manufactured poly(lactic acid) (PLA)/multi-walled carbon nanotube (MWNT) composite filaments through melt spinning with different take-up velocities to obtain electrically conductive filaments. The researchers spun the filaments using a self-constructed piston-type spinning device, where the granular composite material was melted in a heated cylinder and pressed through a single hole die of 0.6 mm in diameter by a piston. The design of the system is illustrated in Fig. 4.5 (Pötschke P *et al.*, 2010).

The volume throughput was controlled by an adjustable piston speed and was kept constant at 0.755 cm^3/min. The melt-spun monofilaments were collected on a winder. The take-up velocity was set to 20, 50, and 100 m/min. The researchers found that the liquid sensing properties of the filaments were strongly influenced by both the MWNT content and the take-up velocity. Both yielded decreased relative resistance changes. It is interesting to note that multi-walled carbon nanotubes had reasonably good macroscopic dispersion and distribution within the PLA as evidenced by the microscopy image shown in Fig. 4.6 (Pötschke P *et al.*, 2010).

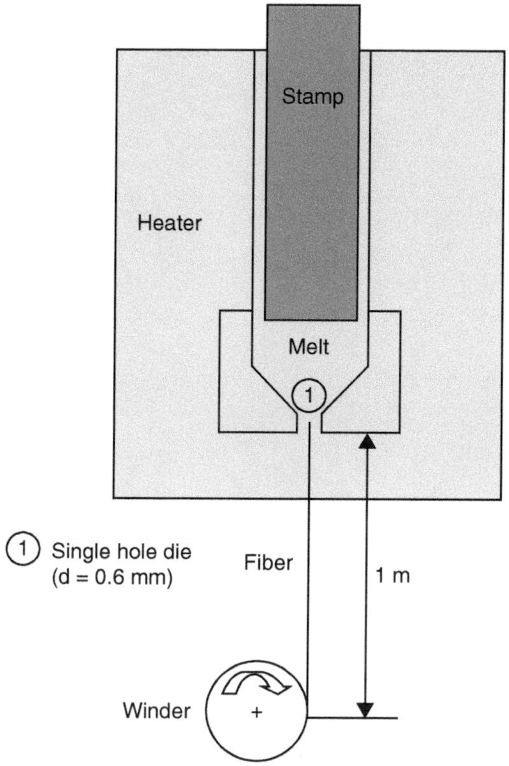

4.5 Piston-type spinning device for production of monofilaments (Pötschke P *et al.*, 2010)

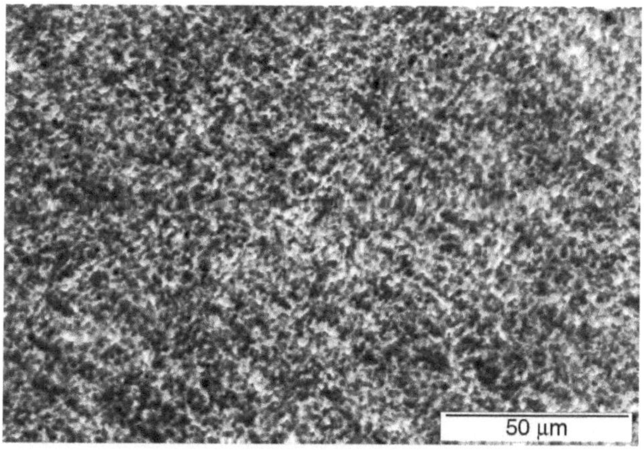

4.6 Light transmission microscopy image of a PLA fiber with 3.0 wt.% MWNT (as extruded) illustrating the high degree of macroscopic MWNT dispersion (Pötschke P *et al.*, 2010).

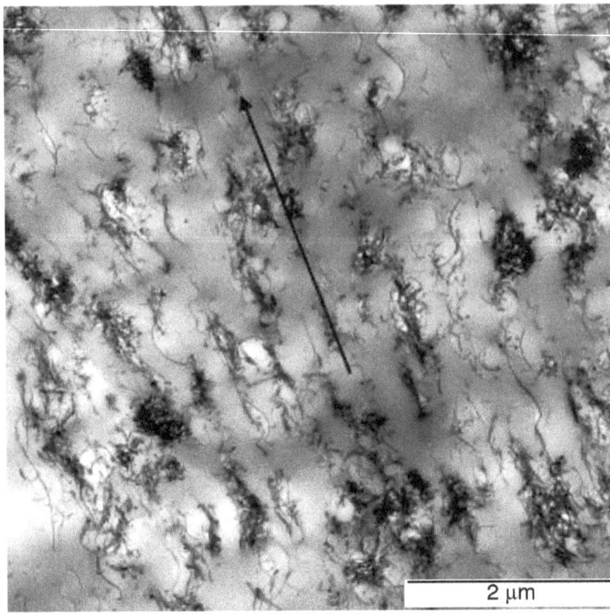

4.7 TEM images of PLA fiber containing 3.0 wt.% MWNT, melt spun with take-up velocity of 50 m/min (arrow indicates fiber axis) (Pötschke P *et al.*, 2010).

TEM images as shown in Fig. 4.7 (Pötschke P *et al.*, 2010) indicated that MWNT are strongly oriented in the filament direction due to the spinning process. The extruded filaments with low MWNT concentration (2 wt.%) and low DDR (extruded) seem to be most suitable for use as sensor filaments in textiles applied in tanks.

In a study conducted by Anand *et al.* (2010), poly(ethylene terephthalate)-single-walled carbon nanotube (PET-SWNT) nanocomposite filaments were prepared through melt spinning and drawing. The researchers found that the nanotubes, in addition to acting as effective nucleating agents for PET melt crystallization, also provided significant reinforcement to the PET filaments. Tenacity and initial modulus of the composite filament with 1 wt.% SWNTs were, respectively, 1.8 times and 3.2 times higher as compared to those of the pristine PET filament prepared under identical conditions. When the draw ratio was increased, the tenacity and the modulus of the filaments further increased, indicating that drawing induced orientation of the polymer molecules and SWNTs. It should be noted that the thermal stability of the filaments was not significantly affected by the presence of the SWNTs at low concentrations.

4.6.3 Structure–property relationship in syndiotactic polypropylene

The crystallization behavior of syndiotactic polypropylene (sPP) was investigated during melt spinning by Sura *et al.* (2001). It was found that its behavior was substantially different from that of most other linear polymers. The anisotropic stress field in the process led invariably to extension as well as alignment (orientation) of the chains in the melt, both of which contributed to a dramatic enhancement in the rate of crystallization. However, the authors commented that the primary structure of the sPP chain in its most preferred crystal form comprised a 'coiled helical,' sequence. Stress-induced chain extension can lead to conformational sequences not favorable for crystallization in this form. As a consequence, process conditions that generate higher stress levels can cause a diminution in the rate of crystallization of this polymer. Such conformation-related aspects of oriented crystallization of sPP have been analyzed in terms of the structure and properties of melt-spun filaments, produced over a range of spinning speeds.

4.6.4 Advance process characterization in melt spinning

Liang and Zhong (2010) measured the elongational properties of low density polyethylene (LDPE) melt using a melt spinning technique. The relationship between temperature and melt strength, as well as between extensional viscosity and drawability, were investigated with respect to extensional strain rate, temperature and extrusion velocity. The researchers found that melt strength decreased with a rise in temperature. A low melt extensional viscosity might be beneficial to improve the melt drawability. With an increase of extensional strain rate, the extensional stress increased while the melt extension viscosity has reduced. Both the melt extensional stress and viscosity dropped with a rise of temperature.

The effect of fluid viscoelasticity on the draw resonance dynamics of melt spinning was investigated by Lee *et al.* (2001) using the White Metzner and Phan-Thien–Tanner fluid models dealing with the effects of spinline cooling on the same draw resonance. The researchers found that the effects of fluid viscoelasticity on spinning stability can be classified into two diametrically different types: for extension-thickening fluids an increase in viscoelasticity increases tension, decreases tension sensitivity and, thus, stabilizes the spinning, whereas for extension-thinning fluids it decreases tension, increases tension sensitivity and, thus, destabilizes the spinning.

In the work by Wu *et al.* (2000), high molecular weight poly(ethylene-2,6-naphthalate)(PEN) was melt spun at take-up velocities between 0.9 and 10 km/min. With increasing take-up speed, a steady trend towards higher as-spun filament orientation and crystallinity was observed, accompanied by improved physical properties. The wide-angle X-ray diffraction (WAXD) patterns of the as-spun filaments prepared at a velocity range higher than 1.5 km/min indicate that these samples all possessed a crystalline structure. At a relatively low take-up velocity, in the range from 1.5 to 4 km/min, a high level of molecular orientation in both crystalline and amorphous regions was observed. This may be attributed to the high spinning stress generated by the high molecular weight polymer used. In the high take-up speed region of 5–10 km/min, the molecular orientation becomes saturated. The highest initial modulus of the as-spun PEN filaments obtained in this region reached 200 g/den. It is interesting to note that the researchers used materials of different molecular weights. The higher molecular weight sample

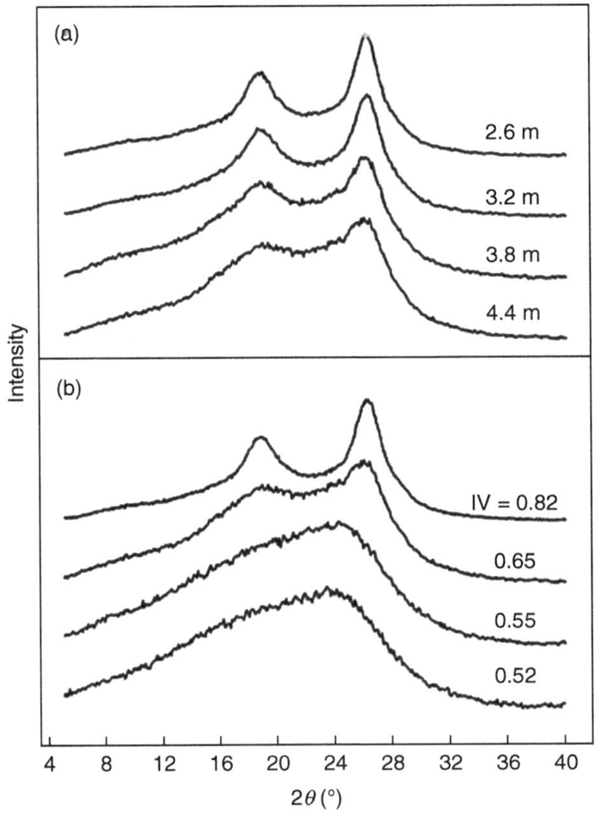

4.8 Equatorial X-ray diffraction profiles of as-spun PEN fibers as a function of: (a) threadline length; (b) molecular weight of polymer (Wu *et al.*, 2000). Here IV represents intrinsic viscosity.

developed higher orientation and crystallization, resulting in well-resolved X-ray diffraction curves (see Fig. 4.8, Wu *et al.*, 2000) and higher values of birefringence, density, and mechanical properties.

In the experimental and theoretical studies of PET melt spinning, it has been found that in the range of high velocities, the inertial stress related with the winding speed plays an important role in the formation of filament structure. In the range of lower velocities, threadline tension is more controlled by the rheological forces. The researchers observed that higher spinning speed generates a higher inertial force, resulting in a significant increase in the threadline tension, and ultimately in high birefringence. At same take-up speed, the rheological drag was enhanced by the use of high molecular weight PEN. The enhanced rheological drag also induced an increased threadline tension, which is considered an important factor in explaining the well-developed structures formed in filament samples wound at quite low speed.

4.7 Future trends

Current developments are focused on obtaining nano- or submicron-filaments using a melt-electrospinning technique. This technique requires the generation of a strong electric field between the polymeric melt within the extruder and a metallic collecting device. The principle is similar to that of electrospinning of filaments from solution; however, the forces needed to create a reduction in diameter to the nanometer level are of great interest. Furthermore, utilization of new polymers and multicomponent spinning, specifically bicomponent systems, are expected to grow in the future. Spinning of new polymers would present new challenges, and further developments are envisaged in the field of post spinning operations such as quenching and drawing. Improvement of the filament quality at high spinning speeds, integration of quality assurance devices with the fast production speeds, and ensuring stability in spinning of finer filaments are some of the research challenges for the years ahead.

4.8 Conclusion

Melt spinning is among the most versatile and commonly used methods for producing polymeric filaments. The fundamentals of melt spinning, modeling of polymeric flows inside the screw and the spinneret, interaction of melt-spun filaments with the environment, and the limitations of melt spinning have been critically discussed. In addition, some of the recent developments and future trends pertaining to melt spinning have been analyzed.

4.9 Sources of further information and advice

Ziabicki A and Jarecki L (1985), *High-Speed Fiber Spinning*, New York, Interscience.

Tucker, C L (1989), *Fundamentals of Computer Modeling for Polymer Processing*, Munich, Carl Hanser Verlag.

Salem, D R (2001), *Structure Formation in Polymeric Fibers, Munich*, Carl Hanser Verlag.

Nakajima, T (1994), *Advanced Fiber Spinning Technology*, Cambridge, Woodhead Publishing Limited.

McIntyre, J E (2005), *Synthetic Fibres: Nylon, Polyester, Acrylic, Polyolefin*, Cambridge, Woodhead Publishing Limited.

Deopura B L, Alagirusamy R, Joshi M, Gupta B (2008), *Polyesters and Polyamides*, Cambridge, Woodhead Publishing Limited.

4.10 References

Anand K A, Jose T S, Agarwal U S, Sreekumar T V, Banwari B and Joseph R (2010), 'PET-SWNT nanocomposite fibers through melt spinning', *International Journal of Polymeric Materials*, **59**, 438–449.

Arai T and Aoyam H (1963), 'Die wall restriction on elastic shear deformation in viscoelastic flow of polymer melt', *Transactions of Society of Rheology*, **7**, 333–355.

Arai T and Toyoda H (1970), 'Viscoelastic flow through a rectangular duct', *Proceedings of 5th International Rheological Congress*, Kyoto, 7–11 October 1968, 461–470.

Bagley E B, Storey S H and West D C (1963), 'Post extrusion swelling of polyethylene', *Journal of Applied Polymer Science*, **7**, 1661–1672.

Beyreuther R and Brunig H (2007), *Dynamics of Fiber Formation and Processing*, Berlin Heidelberg, Springer-Verlag.

Bhat G S, Malkan S R (2007), 'Polymer-laid web formation', in Russell S J (ed.), *Handbook of Nonwovens*, Cambridge, Woodhead Publishing Limited.

Bird B R (1977), *Dynamics of Polymeric Liquids*, New York, John Wiley and Sons.

Booy M L (1963), 'Influence of channel curvature on flow, pressure distribution and power requirement of screw pumps and melt extruders', *SPE Transactions*, **3**, 176–185.

Buchdahl R, Nielson L E and Merz E H (1951), 'Rheology of thermoplastics. II. Recoverable and nonrecoverable deformations of polystyrene between 90°C and 240°C, as a function of molecular weight', *Journal of Polymer Science*, **6**, 403–422.

Campbell G A, Sweeney P A and Felton J N (1992), 'Experimental investigation of the drag flow in extruder analysis' *Polymer Engineering and Science*, **32**, 1765–1770.

Chang P W, Patten T W and Finlayson B A (1979), 'Collocation and galerkin finite element methods for viscoelastic fluid flow—I.: Description of method and problems with fixed geometry', *Computers and Fluids*, **7**, 267–283.

Chin W Y and Shyu G D (1988), 'Die swell from capillary die and slit die: A theoretical study', *Journal of Applied Polymer Science*, **35**, 847–862.

Choi D and White J L (2000), 'Structure development in melt spinning syndiotactic polypropylene and comparison to isotactic polypropylene', *International Polymer Processing*, 2000/04, **15**, 398–405.

Choo K P, Neelakantan N R and Pittman J F T (1980), 'Experimental deep-channel velocity profiles and operating characteristics for a single-screw extruder', *Polymer Engineering Science*, **20**, 349–356.

Dhadwal R (2011), 'Numerical simulation of a two-phase melt spinning model', *Applied Mathematical Modelling*, **35**, 2959–2971.

Duda J L and Vrentas J S (1972), 'Pressure losses in non-Newtonian entrance flows', *The Canadian Journal of Chemical Engineering*, **50**, 671–674.

Fourné F (1999), *Synthetic Fibers: Machines and Equipment, Manufacture, Properties*, Munich, Carl Hanser Verlag.

Graessley W W, Glasscock S D and Crawley R L (1970) 'Die swell in molten polymers', *Transactions of Society of Rheology*, **14**, 519–544.

Hooshmand S, Soroudi A and Skrifvars M (2011), 'Electro-conductive composite fibers by melt spinning of polypropylene/polyamide/carbon nanotubes', *Synthetic Metals*, **161**, 1731–1737.

Jeon B S (2001), 'Theoretical orientation density function of spunbonded nonwoven fabric', *Textile Research Journal*, **71**, 509–513.

Jinan C, Kikutani T, Takaku A and Shimizu J (1989), 'Nonisothermal orientation-induced crystallization in melt spinning of polypropylene', *Journal of Applied Polymer Science*, **37**, 2083–2097.

Kantorovich L V and Krylov V I (1958), *Approximate Methods of Analysis*, New York, Interscience.

Kase S and Matsuo T (1965), 'Studies on melt spinning-I, Fundamental equations on the dynamics of melt spinning', *Journal of Polymer Science*, **A3**, 2541–2554.

Kim J Y and Kim S H, (2006), 'Structure and property relationship of thermotropic liquid crystal polymer and polyester composite fibers', *Journal of Applied Polymer Science*, **99**, 2211–2219.

Lee J S, Jung H W, Kim S H and Hyun J C (2001), 'Effect of fluid viscoelasticity on the draw resonance dynamics of melt spinning', *Journal of Non-Newtonian Fluid Mechanics*, **99**, 159–166.

Li Y and Hsieh F (1996), 'Modelling of flow in a single screw extruder', *Journal of Food Engineering*, **27**, 353–373.

Liang J Z and Zhong L (2010), 'Elongational properties of low density polyethylene using meltspinning technique', *Polymer Testing*, **29**, 972–976.

Lu F M and Spruiell J E (1987), 'The influence of resin characteristics on the high speed melt spinning of isotactic polypropylene I: Effect of molecular weight and its distribution on structure and mechanical properties of as-spun filaments', *Journal of Applied Polymer Science*, **34**, 1521–1539.

Mayer H J, Stiehl C and Roeder E (1997), 'Applying the finite-element method to determine the die swell phenomenon during the extrusion of glass rods with non-circular cross-sections', *Journal of Materials Processing Technology*, **70**, 145–150.

Metzner A B, Houghten W T, Sailor R A and White J L (1961), 'A method for the measurement of normal stresses in simple shearing flow', *Transactions of Society of Rheology*, **5**, 133–147.

Middleman S (1965), 'Flow of polymer fluids in rectangular ducts', *Transactions of Society of Rheology*, **9**, 83–93.

Mitsoulis E, (1999), 'Three-dimensional non-Newtonian computations of extrudate swell with the finite element method', *Computer Methods in Applied Mechanics and Engineering*, **180**, 333–344.

Mitsuishi N and Aoyagi Y (1969), 'Non-Newtonian flow in non-circular ducts', *Chemical Engineering Science*, **24**, 309–319.

Mitsuishi N and Aoyagi Y (1973), 'Non-Newtonian flow in eccentric annulus', *Journal of Chemical Engineering*, **6**, 402–408.

Mori and Funatsu (1973), 'On die swell in molten polymer', *Journal of Applied Polymer Science*, **20**, 209–220.

Nickell R E, Tanner R I and Caswell B (1974), 'The solution of viscous incompressible jet and free-surface flows using finite-element methods', *Journal of Fluid Mechanics*, **65**, 189–206.

Pötschke P, Andres T, Villmow T, Pegel S, Brünig H, Kobashi K, Fischer D and Häussler L (2010), 'Liquid sensing properties of fibers prepared by melt spinning from poly(lactic acid) containing multi-walled carbon nanotubes', *Composites Science and Technology*, **70**, 343–349.

Racin R and Bogue D C (1979), 'Molecular weight effects in die swell and in shear rheology', *Journal of Rheology*, **23**, 263–280.

Rawal A (2002), 'Generation of an expert system for the optimisation of net-extrusion processes', Ph.D Thesis, The University of Bolton (*formerly* Bolton Institute), Bolton, UK.

Rawal A and Davies P J (2005), 'Expert system for the optimisation of melt extruded net structures', *Plastics, Rubber and Composites*, **34**, 47–53.

Rawal A and Davies P J (2006), 'Empirical Modelling of Filament Area in Melt Extruded Net Structures', *Indian Journal of Fiber and Textile Research*, **31**, 241–247.

Rejon A and Alvarez C (1987), 'Mechanical and flow properties of high density polyethylene/low density polyethylene', *Polymer Engineering Science*, **27**,640–646.

Richardson S (1970), 'The die swell phenomenon', *Rheologica Acta*, **9**, 193–199.

Rowell H S and Finlayson D (1922), 'Screw viscosity pumps', *Engineering*, **114**, 606–607.

Rowell H S and Finlayson D (1928), 'Screw viscosity pumps', *Engineering*, **126**, 249–387.

Rwei S P (2000), 'Modification of PET in high-speed melt spinning by blending with PEN', *Polymer Engineering and Science*, **40**, 191–200.

Schechter R S (1961) 'On the steady flow of a non-Newtonian fluid in cylinder ducts', *AIChE Journal*, **7**, 445–448.

Sumesh P T, Mathur T P and Agarwal U S (2010), 'Simulation of polyester melt spinning with axial quench for increasing productivity', *Journal of Applied Polymer Science*, **116**, 2541–2547.

Sura R K, Desai P and Abhiraman A S (2001), 'Oriented crystallization in fiber formation: Inferences from the structure and properties of melt spun syndiotactic polypropylene filaments', *Journal of Applied Polymer Science*, **81**, 2305–2317.

Sutterby J L (1966), 'Laminar converging flow of dilute polymer solutions in conical sections: Part I. Viscosity data, new viscosity model, tube flow solution', AIChE *Journal*, **12**, 63–68.

Tadmor Z and Gogos C G (1979), *Principles of Polymer Processing*, New York, John Wiley and Sons Inc.

Takarada W, Ito H, Kikutani T and Okui N (2001) 'Studies on high-speed melt spinning of noncircular cross-section fibers I: Structural analysis of as-spun fibers', *Journal of Applied Polymer Science*, **80**, 1575–1581.

Tanner R I (1970), 'A theory of die-swell', *Journal of Polymer Science*, **8**, 2067–2078.

Tanner R I, Nickell R E and Bilger R W (1975), 'Finite element methods for the solution of some incompressible non-Newtonian fluid mechanics problems with free surfaces', *Computer Methods in Applied Mechanics and Engineering*, **6**, 155–174.

Treloar L R G (1958), *The Physics of Rubber Elasticity*, Oxford, Clarendon Press.

Vlachopoulos J, Horie M and Lidorikis S (1972), 'An evaluation of expressions predicting die swell', *Transactions of Society of Rheology*, **16**, 669–685.

White J L and Roman J F (1976), 'Extrudate swell during the melt spinning of fibers—influence of rheological properties and take-up force', *Journal of Applied Polymer Science*, **20**, 1005–1023.

Wong A C Y and Liang J Z (1997), 'Relationship between die swell ratio and melt flow index', *Chemical Engineering Science*, 52, 3219–3221.

Worth R A (1979), 'Accuracy of parallel – plate analogy for representation of viscous flow between cylinders', *Journal of Applied Polymer Sciences*, **24**, 319–328.

Wu G, Li Q and Cuculo J A (2000), 'Fiber structure and properties of poly(ethylene-2,6-naphthalate) obtained by high-speed melt spinning', *Polymer*, **41**, 8139–8150.

Xue S C, Phan-Thien N and Tanner R I (1995), 'Numerical study of secondary flows of viscoelastic fluid in straight pipes by an implicit finite volume method', *Journal of Non-Newtonian Fluids Mechanics*, **59**, 191–213.

Yarin A L, Sinha-Ray S and Pourdeyhimi B (2010), 'Meltblowing: II-linear and nonlinear waves on viscoelastic polymer jets', *Journal of Applied Physics*, **108**, 034913-1-034913-10.

Yarin A L, Sinha-Ray S and Pourdeyhimi B (2011), 'Meltblowing: Multiple polymer jets and fiber-size distribution and lay-down patterns', *Polymer*, **52**, 2929–2938.

Yu Q and Hu G H (1998), 'Development of a helical coordinate system and its applications to analysis of polymer flow in screw extruders. Part II: A helical channel model for single screw extruders', *Polymer Engineering and Science*, **38**, 819–830.

Ziabicki A and Kedzierska K (1960), 'Mechanical aspects of fiber spinning process in molten polymers Part II. Stream broadening after the exit from the channel of spinneret', *Kolloid Zeitschrift*, **171**, 111–119.

Ziabicki A (1967), 'Physical fundamentals of the fiber spinning process' and 'Principles of melt spinning', in Mark H F, Atlas S M and Cernia E, *Man-Made Fibers Science and Technology Volume* 1, Interscience Publishers, John Wiley Sons, Inc.

Ziabicki A (1976), *Fundamentals of Fiber Formation*, New York, Interscience.

Ziabicki A, Jarecki L and Wasiak A (1998), 'Dynamic modelling of melt spinning', *Computational and Theoretical Polymer Science*, **8**, 143–157.

5
Gel spinning of synthetic polymer fibres

C. J. KUO and W. L. LAN, National Taiwan
University of Science and Technology, Taiwan

DOI: 10.1533/9780857099174.2.100

Abstract: With increasing requirements for better performance of fibre products, the development of reinforced fibre has become a key issue. Gel spinning technology is one of the major production methods for high-strength and high-mechanical-property fibres. Gel spinning, also known as semi-melt spinning, is a method that prepares high-strength and high-elastic modules fibre in the gel state. After the extrusion of the polymer solution or plasticised gel from the spinnerets, it is cooled in solvent or water before being stretched into gel fibre by ultra-high extension. Gel spinning can promote the polymer of fibre's molecular weight, degree of orientation, crystallinity and fibre density effectively, and that contributes to increasing the fibre's strength and modules. In the near future, gel spinning will be able to produce many more special fibres with excellent properties.

Key words: gel spinning, high magnification stretching, high-strength fibre.

5.1 Introduction

Fibre is one of the most important inventions in human history. In early times, people planted cotton and ramie to extract fibre for making clothing to replace the use of leather, even though its quality was often crude and unrefined. The Chinese extracted silk to improve the quality of their clothing; however, the material was very expensive. The invention of rayon at the end of the nineteenth century triggered an upsurge in fibre development. In 1935, the first synthetic fibre was invented and its products were officially marketed in 1940, promoting the rapid development of the fibre industry. To date, the most commonly used spinning methods are melt spinning, dry spinning, and wet spinning.

With increasing requirement for better performance of fibre products, the development of reinforced fibre has become a key issue. In the 1970s, DuPont of the US successfully developed the aramid fibre, with a strength about 17 cN/dtex, thus leading to high-strength fibre development. The main types of high-strength fibres include para-aramid (PPTA) fibre,

Table 5.1 A comparison of fibre strength and modulus

Fibre type	Strength (cN/dtex)	Modulus (cN/dtex)	Process method	Commodity description
PPTA fibre	16–24.5	355–880	Crystal spinning	Kevlar®, Twaron®
UHMWPE fibre	26.5–35	885–1760	Gel spinning	Dyneema®, Spectra®
Carbon fibre	13.5–35	1060–1675	Wet spinning	Torayca®

Source: Shen, 1979 and Yasuda, 1994.

UHMWPE fibre and carbon fibre. A comparison of fibre strength and modulus is shown in Table 5.1.[1,2]

However, with rising demands on fibre product performance, many special spinning methods have been researched and developed. Methods including gel spinning, electrostatic spinning and integrate composite spinning have been applied in industrial production. Gel spinning has been widely applied in the production of high-strength fibres and those with high mechanical properties.[3,4] The first patent of gel spinning emerged in the 1950s, but since the molecular weight of the raw material used at that time was too low, the obtained fibre did not have the expected strength and mechanical properties. As a result, there was no further development. Ten years later, the development of carbon fibre of sufficiently high strength and high mechanical property has put gel spinning in the spotlight once again. Gel spinning technology has been continuously upgraded and improved, and its range of application has extended from the PE of low molecular weight to raw materials of high molecular weight, such as polyvinyl alcohol (PVA), polyacrylonitrile (PAN) and ultra-high molecular weight polyethylene (UHMWPE). Gel spinning technology has become one of the major production methods of high-strength and high-mechanical-property fibres.

5.2 Factors affecting the spinning of polymers: entanglement

Entanglement refers to the reticular or spherical structure formed by the cross-linking points produced in the polymer chain, or in between polymer chains, making the polymer chains unable to move normally and thus affecting the nature of the polymer. The measurement of entanglement is generally determined by the viscosity of the polymer fluid or the modulus of complete non-crystalline polymer.

Entanglement has a significant impact on the rheological behaviour and mechanical properties of the polymer. Observed from the perspective of polymer melt viscosity, when the shear rate is at zero the polymer viscosity is proportional to the molecular weight of the polymer. When the molecular weight exceeds a certain level, the viscosity is proportional to the 3.4th power of the molecular weight as shown in Equations [5.1] and [5.2]:

$$\eta_0 \propto M \ (M < M_c) \quad [5.1]$$

$$\eta_0 \propto M^{3.4} \ (M \geq M_c) \quad [5.2]$$

where M_c is the critical entanglement molecular weight, which is generally in the range of 10 000–20 000. When the molecular weight exceeds M_c, the viscosity increases significantly. Thus, it proves the existence and impact of entanglement on the physical properties of the polymer molecule.

Polymer chain entanglement has a significant effect on the processing of polymer fibre. According to rubber elasticity theory, the maximum stretch ratio of polymer fibre (λ_{max}) and the number of statistical links of the polymer chain cross-linking points are in the relationship shown in Equation [5.3]. When the number of statistical links is greater, the amount of stretch after fibre processing will be greater. In the ideal state, when the polymer chains are completely parallel in arrangement, the amount of stretch will be infinite, and the fibre strength will accordingly be infinite. Hence, if the number of polymer entanglements can be effectively reduced, the polymer fibre strength will be considerably improved.

$$\lambda_{max} = \sqrt{N_C} \quad [5.3]$$

5.3 Factors affecting fibre strength

Like cement and reinforced steel bars, for the preparation of high-strength polymer fibre the internal structure of the fibre is important. The preparation of high-strength polymer is now discussed from different perspectives, including the polymer molecular weight, degree of orientation, crystallinity, fibre density and processing method.

5.3.1 Molecular weight

Molecular weight is one of the most important considerations for preparing high-strength fibre. Rising molecular weight can effectively reduce the defects at the molecular chain ends to enhance fibre strength. Rising molecular weight can also improve the maximum stretch ratio, and further affect the molecular phase smoothness and crystallinity. However, rising molecular weight relatively raises the polymer melt or solution viscosity to reduce the spinnability of the fibre.

5.3.2 Degree of orientation and crystallinity

The degree of orientation and crystallinity of the polymer can directly affect polymer fibre strength. A good degree of orientation and crystallinity can enable the molecules to bear greater stress. The high degree of orientation of polymer structure is related to the molecular composition or the characteristics of medium solution. Different molecular or solution compositions will result in different adaptive structural patterns. In addition to the selection of molecule and solution, special processing methods, such as the sudden cooling of melt fibre, fibre ultra-magnification stretching and multi-segment stretching of the fibre, can improve the degree of orientation and crystallinity.

5.3.3 Fibre density and processing methods

Fibre density directly affects its strength. Polymer molecular weight, solution concentration, fibre processing methods and fibre stretching affect fibre density directly. Melt spinning without solution can theoretically achieve the maximum density of the fibre. However, given the need to ensure spinnability of the fibre, the melt spinning process generally uses polymers of lower molecular weight, which does not produce high-strength fibres. The dry spinning method also obtains a similar fibre density; however, the stretching effects are not significant. Limited by the problem of swell, the wet spinning method does not produce high-strength fibre.

The liquid-crystal spinning method can also be used to produce high-strength fibre. However, the molecules need to be rigid-chain polymers that can be made into liquid solution. The gel spinning method may solve the polymer entanglement problem and improve the spinnability of polymer of higher molecular weight. Coupled with ultra-magnification stretching, the molecular degree of orientation and crystallinity can be increased to effectively improve the fibre strength. Therefore, gel spinning is one of the major processing methods for making high-strength fibres.

5.4 Gel spinning technologies

Gel spinning, also known as semi-melt spinning, is a method that prepares high-strength and high-elastic module fibre in the gel state. After the extrusion of the polymer solution or plasticised gel from the spinnerets, it is cooled in solvent or water before being stretched into gel fibre by ultra-high extension. During the process of cooling, the polymer solution will gradually lose the mobility of the solution, which is known as the gelation of polymers. During the process of gelation, the solution's viscosity will gradually rise. Until the formation of gel, the viscosity will become infinite. As the new forming fibre contains a large amount of solvent, and is in the gel state, it requires ultra-drawing to produce the ultra-high-strength and high-modulus fibre. Gel spinning causes relatively ideal structures for ordinary flexible-chain polymers (e.g. PE, PAN and PVA) to produce high-strength and high-modulus fibre.[5]

The general gel spinning process can be divided into the following four steps:

1. Dissolution: the dissolution of polymer in solvent. The molecular chains of the solid polymers may be entangled, and the entanglements will affect the molecular chain stretching. The dissolution of polymers in solvent can disentangle these entanglements to aid the gel fibre stretching.
2. Spinning and formation: extruding the solvent from the spinnerets by gear pump, and cooling it rapidly by air or water. The rapid cooling process can lead to the formation of crystals and retain the disentangled state of polymers to aid the formation of high-strength and high-elastic-modulus fibre.
3. Remove solvent: the most commonly used methods to remove solvent in gel spinning include the natural drying method and the use of extractant, which mainly aims to remove the solvent residuals in gel fibre. The extractant can displace the gel fibre solvent based on diffusion and penetration principles. In gel spinning, the selection of solvent and extractant directly affect the stretching stability, and is key to the gel spinning method.
4. Stretching: the high magnification stretching of the gel fibre. High magnification stretching can change the folded polymer molecular chains to straight chains, and improve polymer crystallinity and orientation. This is a unique feature of gel spinning, and is the key to enhancing gel fibre performance.

Using solvent and high magnification stretching, the gel spinning process can produce gel fibre of high strength and high Young's modulus. The process is simple and similar to the wet and dry spinning processes, with the following differences:

1. Raw materials of high molecular weight: the raw materials are mainly polymers of high molecular weight. Polymers of high molecular weight

can withstand high magnification stretching, and help improve fibre strength and other mechanical properties. The rising molecular weight can reduce defects in fibre structure caused by the chain polymer end group to enhance fibre strength.

2. Dilute solution: the diluted solution can reduce gel viscosity, improve gel fibre spinning capability, and be conducive to the disentanglement of high polymer chains to enhance fibre strength and mechanical properties.
3. High magnification stretching: as stated above, the high magnification stretching of fibre is a unique feature of the gel spinning process. The high magnification stretching can change the folded polymer molecular chains to straight chains, and improve polymer crystallinity and orientation, which is an incomparable advantage over other spinning methods.

5.5 Types of fibres produced using gel spinning

Gel spinning is usually applied in the processing of high molecular weight or ultra-high molecular weight polymers. The entanglement of long chain polymers is fully stretched by appropriate solvent and temperature. After spinning, high magnification stretching can further stretch the polymer molecular chain to get the high strength of the fibre. Gel spinning is most widely applied in production of UHMWPE, PVA and PAN, as illustrated respectively below.

5.5.1 UHMWPE fibre

UHMWPE uses molecules of high molecular weight as the raw material, because increasing molecular weight can reduce the number of end groups and subsequent end group defects of the molecular chain to improve fibre strength. At present, the gel spinning of UHMWPE uses solvents including paraffin oil, decalin and paraffin.[6,7] Decalin has the best dissolution effect; however, it cannot be mass-produced due to its high price. At present, a mixture of kerosene and emulsifying oil is usually used as the solvent. The most commonly used extractants are trichloro-trifluoro-ethane, decane, gasoline, and the high efficiency dragging oil smoke. Stretching UHMWPE fibres uses the high magnification stretching method to straighten the folded molecular chain to the largest extent to enhance UHMWPE's mechanical properties. UHMWPE fibre has many excellent mechanical properties and high energy absorption value, and thus is usually used in producing helmets, impact plates and bullet-proof vests.

5.5.2 PVA fibre

PVA fibre has good heat resistance and adhesion with rubber as well as excellent mechanical properties. Using the gel spinning method to prepare high-strength and high-mechanical-property fibres has become an integral part of the development of PVA products. PVA gel spinning is not easy. The spinning requires special technologies and methods, such as a low temperature coagulation bath, a two-stage reduction of tension extraction drying, and high tensile heat setting in order to produce a PVA fibre of high strength and high modulus.[8] In applications, PVA fibre is often used as the raw material for nylon.

5.5.3 PAN fibre

PAN fibre gel spinning process usually uses acrylonitrile homopolymer or copolymer of high molecular weight as the raw material. Gel spinning is conducted by 15 wt.% dilute solution. The solvents include NaScN water solution, DMF, DMSO, DMAC, and δ-butyrolactone, and ultra-high magnification stretching of gel fibre is implemented. The PAN fibre obtained from the gel spinning process has few defects and high orientation with mechanical properties measurable to aromatic polyamide fibre. Regarding thermal stability, PAN fibre is worse than aromatic polyamide fibre, carbon fibre, and glass fibre, but is considerably better than PE fibre. The strength is not affected in environments above 130°C.[9] Moreover, PAN fibre has excellent anti-alkaline properties and light fastness.

5.6 Factors affecting gel spinning

Many factors affect gel spinning processing. The product's spinning property, stability and quality are subject to factors such as polymer type, molecular weight, solvent type, solvent concentration, laddering rate, spinning temperature, extractant type, stretching temperature, stretching rate, and stretching magnification, as summarised in Fig. 5.1

1. Molecule type: molecule type determines the use of the product. The type of molecule chosen for the gel spinning process is usually polymers of chain thermo-plasticity, and commonly used gel spinning molecules include UHMWPE, PVA and PAN.
2. Molecular weight: the polymer molecular weight determines the fibre strength and modulus after gel spinning. The molecular weight directly affects the length of the molecular chain. Usually, longer molecular chains will have higher fibre strength and modulus.

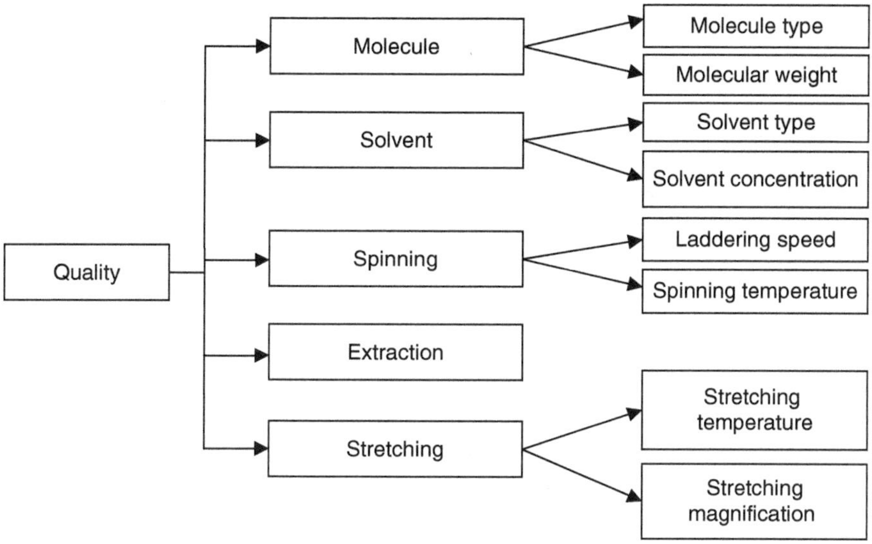

5.1 Factors affecting gel spinning.

3. Solvent type: selection of solvent is highly important to gel spinning. The dissolution rate of polymers directly affects the spinning property and stability of the gel fibre.
4. Solution concentration: solution concentration directly affects the maximum stretching magnification of the gel fibre; however, the impact is usually subject to the molecule type. Taking PAN as an example, PAN gel spinning usually occurs at low concentration. The lower concentration can lead to higher maximum stretching magnification of the gel fibre and higher fibre modulus.
5. Ladder speed: in general, higher ladder speed can result in better fibre extensibility. However, excessively high ladder speed often leads to excessive shearing rate between the fluid and the wall of the flow channel. As a result, the molecular chain alleviating speed cannot keep up with the spinning speed to increase surface stress and internal defects of fibre. In addition, in the high stretching magnification of polymers, excessively high ladder speed can cause slippage between fibrillar and microfibrillar to lead to plastic deformation of the fibrillar inside the fibre. As a result, under the external force, the fibrillar will be torn along the axial direction of the microfibrillar to reduce the fibre strength. Hence, appropriate ladder speed is an important factor that affects the quality of gel spinning products.
6. Spinning temperature: in theory, a higher spinning temperature will make the dissolution of the spinning solution more even. However,

relatively high temperature can easily lead to the thermal cracking of the fibre and reduce the viscosity of the spinning solution. Hence, the entanglements of the molecular chains will be reduced, and the arrangements will be loosened to increase the fibre defects. In general, the gel spinning temperature is around the polymer single crystal dissolution temperature, resulting in better extension effects. Appropriate spinning temperature is also an important factor that affects the quality of gel spinning products.

7. Extraction: extraction of gel fibre plays an important role in gel spinning. Appropriate extractant can reduce the impact of the solvent residuals on the stretching of the gel.
8. Stretching temperature: the gel spinning stretching temperature directly affects the maximum gel stretching magnification, and indirectly affects the modulus of the gel fibre. The rising temperature of gel spinning stretching can promote the mobility in between single molecular chains to improve the extensibility of the gel fibre until reaching the melting point of the polymer. Hence, the setting of the stretching temperature is usually slightly below the melting point.
9. Stretching magnification: from the perspective of fibre internal crystallinity, orientation and crystal morphology, under low-magnification extension of polymer gel, the crystals will gradually change into microfibrillar with a sharp increase in orientation and gradually rising strength and crystallinity. When the stretching magnification gradually increases, the growth of orientation tends to slow down while the strength and growth rate of the crystals also reduces. When it reaches half of the maximum stretching magnification, spot defects occur on the surface of the gel fibre. The fibre's internal orientation, strength and crystallisation growth become even slower. When the stretching magnification gradually increases, the molecular chains become gradually intense and the fibre internal crystallisation starts to change to make the arrangement of microfibrillar denser. The rising single monoclinic crystal percentage can also increase the number of molecular chains in the unit area, and the fibre strength and modulus will be enhanced to the limit.[10] When the stretching magnification reaches the maximum, the crystallisation is damaged and will break the fibre. It can be confirmed by molecular internal morphology that the axial extension of gel fibre can get to the state of high strength and high modulus in case of high stretching magnification. Hence, stretching magnification is the most important factor in the gel spinning fibre processing process.

Many factors affect the gel spinning process. The gel fibre can obtain the high strength and high modulus only by properly adjusting these factors.

5.7 Gel spun textile products

Gel spinning fibre is applied in high strength and high tensile fabrics, including bullet-proof vests, belts, gloves and sportswear. The most popular gel spinning fibre, UHMWPE, can be used to manufacture safety protective equipment, high-tension ropes, fishing nets and sports leisure goods as outlined below.

5.7.1 Safety protective equipment

UHMWPE composite fibre has excellent impact resistance performance. It ranks top among composite materials in impact shock absorption, being three times that of carbon fibre and 2.5 times that of aramid.[11,12] Hence, it is suitable for making helmets, body armour, and bullet-proof armour. The composite material of UHMWPE and resin has a light and thin structure with high impact durability and high tensile strength as well as mechanical properties far beyond traditional glass fibre-enhanced composite materials.

The composite material of UHMWPE and resin has been compared with the fibre Kevlar, of the highest strength. In testing of bullet-proof clothing, the composite material of UHMWPE and resin in different fibre stacking and woven patterns can have the same bullet-proof effects as the composite material of Kevlar fibre. Since UHMWPE fibre has a non-polar crystalline arrangement, it has better durability than Kevlar fibre. Regarding long-term use, the bullet-proof functionality of the composite material of UHMWPE fibre is far better than that of the composite material of Kevlar fibre. It is believed that UHMWPE fibre composite material has more development potential than Kevlar fibre composite material for future applications of tensile strength and bullet-proof function. In the body armour market, Kevlar accounts for 48%, while Spectra, which is made from UHMWPE fibre, accounts for 32%. It is believed that the tensile strength and bullet-proof applications of UHMWPE fibre composites have the prospect of exceeding Kevlar fibre composites in the future.

5.7.2 High tensile ropes

UHMWPE fibre is characterised by high strength, high modulus, wear resistance, light fastness and corrosion-resistance. It is suitable for making high-strength ropes, such as cables, ship class ropes and machinery ropes. Compared with the ropes made by aramid, the UHMWPE fibre rope is half the weight and stronger by 10%, with a radius of only 90% of the aramid rope. In addition, the UHMWPE fibre density is maintained under water, and its breaking strength in water is great, making it suitable for ropes in regular contact with water, such as ropes used on oil tankers, in marine engineering and for anchors.

5.7.3 Fishing nets

At present, fishing nets are mostly made of PE. Under the same strength, fishing nets made of UHMWPE are only half the weight of PE fishing nets, and hence can effectively reduce the energy consumption of the fishing boat and improve the speed of the trawl. Therefore, it is suitable for deep-sea fishing.

5.7.4 Sports and leisure goods

UHMWPE fibre has good strength and specific rigidity with toughness and abrasion resistance. It can be used to make rackets, surfboards and snow boards. For its flexibility and light weight, it can be compared with carbon fibre.

5.7.5 Composite materials

UHMWPE can be coupled with PBA-type to improve the resin or coupled with PBA-type acrylic to improve unsaturated polyester composite material in order to enhance the mechanical strength of the original material. Among the synthetic material of PBA-type improved resin and UHMWPE fibres, the PBA-type of polyester has the best mechanical strength improvement effects in terms of tension strength, durability strength, bending strength and shearing strength. The improvement is up to 40%, as compared with original resin material.[13] The synthetic material of PBA-type acrylic improved unsaturated polyesters and UHMWPE fibre can improve impact durability, and reduce the volume shrinkage due to the addition of the PBA-type in the unsaturated polyester. Moreover, with the rising content of PBA-type acrylic, it can improve the functionality and application value of the synthetic material of the unsaturated polyester resin and UHMWPE fibre.

The fibre surface of UHMWPE can be processed by treatment of plasma and chemical etching to achieve surface roughness in order to form better interface adhesion of the synthetic material after surface improvement for high tensile strength.

5.8 Conclusion and future trends

There have been many studies and patents concerning gel spinning. Won-Ill Cha *et al.*[14] used methanol as a solvent and coagulant, and a mixture of dimethyl sulfoxide (DMSO) and water (80/20) to prepare PVA fibre. They found that the PVA fibre prepared under this gel spinning condition could be elongated to 45 times draw ratio, the highest tensile strength attained 2.8 GPa, and the highest Young's modulus attained 64 GPa. Masayosi Suzuki *et al.*[15] prepared atactic and syndiotactic-rich poly(vinyl alcohol) compound fibres by gel spinning. The fibre amorphous parts as a consequence of the difficulty of drawing were considered to form the voids and cracks, which had a banded structure. Xiefei Zhanga *et al.*[16] made single wall carbon nanotubes (SWNT), polyvinyl alcohol (PVA), dimethyl sulfoxide (DMSO) and water, in a homogeneous dispersion which was prepared by spinning. As a result, the compound fibre achieved less crystallinity. Shuiping Liu *et al.*[17] developed a new gel spinning method that was employed to prepare polyacrylonitrile (PAN) fibres. This method achieved a more compact structure, fewer voids and better mechanical properties after a three-stage drawing.

In terms of patents, Schellekens Ronald Michael Ale *et al.* developed a process for the preparation of polyvinyl alcohol articles of high strength and modulus (EP0212757, 1987); Roger Y. Leung prepared silicon carboxide fibres from gel spinning cyclosiloxane polymer precursors (US5629249, 1988); Eric H. M. Hogenboom *et al.* developed a process for manufacturing stretched rope by gel spinning (US5176862, 1991); Sang Heon Kim *et al.* used gel spinning moulding techniques to prepare porous polymer scaffold for tissue engineering (US20070009570, 2006); Joseph Arnold Paul Maria J.A.P.M Simmelink *et al.* developed a new process for spinning UHMWPE (US20100268331, 2008); Joseph Arnold Paul Maria Simmelink developed a new polyethylene fibre by gel spinning (US20110207907, 2009); and Kenneth C. Harding *et al.* used a gel spinning process to manufacture high-modulus ultra-high molecular weight polyethylene tape (US8206810, 2011).

There are many organic high polymers. The successful development of gel spinning has been realised in UHMWPE, PVA, PAN and cellulose ester. The products are mainly made of UHMWPE fibre. UHMWPE fibre has low density, good mechanical performance, good conductivity, impact durability, light fastness and anti-chemical erosion. However, the low melting point and durability of creeping change are the most apparent defects in application. Thus, using the gel spinning method to successfully produce high-mechanical-property nylon or PET fibre will be a great step forward in the spinning field. In short, gel spinning is a unique spinning method with great advantages for producing high-strength and high-modulus fibres.

Overall, the key points concerning the future development of gel spinning are as follows: (1) extension of the molecular chain; (2) reduction of end groups as far as possible; and (3) disentanglement of parts of the molecular chain to separate the entangled molecules to help extension and improve the stretching magnification, as well as to achieve molecular arrangements that can be easily stretched. The equipment for gel spinning is not complex and the technology is relatively simple, and thus it has very wide application prospects. In the near future, gel spinning will be able to produce many more special fibres with excellent properties.

5.9 References

1. Shen S.Y. (1979), *Advanced Polymer Materials*
2. Yasuda H., Ban K. and Ohta Y. (1994), *Advanced Fibre Spinning Technology*, vol. 25, Woodhead Publishing Limited, 172–186.
3. Murase Y. and Nagai A. (1994), *Advanced Fibre Spinning Technology*, vol. 40, Woodhead Publishing Limited, 25–64.
4. Nakagawa J. (1994), *Advanced Fibre Spinning Technology*, vol. 12, Woodhead Publishing Limited, 160–171.
5. Kunugi T. (1998), *J. Appl. Polym. Sci.*, **67**, 1993.
6. Bastiaansen C.W.M. (1992), *Polymer*, **33**(8), 1649–1652.
7. Kunugi T., Kawasumi T. and Ito T. (1990), *J. Appl. Polym. Sci.*, **40**, 2101.
8. Bastiaansen C.W.M. (1990), *J. Appl. Polym. Sci.*, **28**, 1475.
9. Waddon A.J. and Keller A. (1990), *J. Appl. Polym. Sci., Polym. Phys Edn.*, **28**, 1063.
10. Smith Jr K.J. (1990), *Polymer. Eng. Sci.*, **30**, 437.
11. Qian R., Wu L., Shen D., Napper D.H. and Mann R.A. (1993), *Macromolecules*, **26**, 2950–2953.
12. Penning J.P. (1993), *Polym. Bull.*, **31**(2), 242–248.
13. Smith P., Lemstra P.J. and Booij H.C. (1981), *J. Appl. Polym. Sci.*, **19**, 877.
14. Cha W., Hyon S.H. and Ikada Y. (1994), *Polym. Phys.*, **8**, 297.
15. Suzuki M., Tanigami T., Matsuzawa S. and Yamaura K. (1970), *J. Appl. Polym. Sci.*, **8**, 1970–1977.
16. Zhang X., Liu T., Sreekumar T.V., Kumar S., Hu X. and Smith K. (2004), *Polym. Sci*, **8**, 8801.
17. Liu S., Tan L., Pan D. and Chen Y. (2011), *Polym. Sci*, **5**, 453.

6
Bi-component and bi-constituent spinning of synthetic polymer fibres

S. MUKHOPADHYAY, Indian Institute of Technology Delhi, India

DOI: 10.1533/9780857099174.2.113

Abstract: Bi-component melt-spinning has been widely applied to produce functional and novel fibres such as hollow fibres, electrically conductive fibres, etc. This chapter details the extrusion of bi-component fibres and discusses issues affecting the structure-property of the fibres during the spinning process, interfacial interaction parameters and compatibility of the polymers to be chosen for sheath-core spinning. It also details the application areas of bi-component fibres and future directions of research in this field.

Key words: filament, extrusion, bi-component spinning, bi-component, bi-constituent, fibres, synthetic.

6.1 Introduction

Bi-component fibres are a class of polyblend fibres, manufactured by special spinning techniques. The objective of spinning such fibres is to overcome certain limitations encountered during conventional single component spinning. The principal purpose of blending polymers for such specialised spinning is to achieve improved processing and properties for particular end uses. The American Society of Testing Materials (ASTM) defines a bi-component fibre as 'a fibre composed of two physically and/or chemically distinct polymeric components in continuous, longitudinal contact within the fibre'. In keeping with US regulations (1997), the official term for a fibre manufactured from two or more components is *bi-constituent,* although the more common terms *bi-component*, *cofilate* or *cospun* are used in both Europe and the USA, while the term *conjugate fibres* is used in Asia.

Dupont introduced the first commercial bi-component application in the mid 1960s. This was a side-by-side hosiery yarn called 'cantrese' and was made from two nylon polymers which, on retraction, formed a highly coiled elastic fibre. In the 1970s, various bi-component fibres began to be produced in Asia, notably in Japan. Highly complex and apparently expensive spin packs were used in the manufacturing process. These techniques were found to be technically unsatisfactory and excessively expensive. Later, in 1989, a novel approach was developed using thin flat plates with holes and grooves to route the polymers. This process was very flexible and quite cost-effective.

Bi-component fibres can be made from two variants of the same generic fibre (two types of polyester (PES), polyethylene terephthalate (PET), copolyester (coPES), polyacrylonitrile (PAN), polyamide (PA), polyethylene (PE) and polypropylene (PP)). The fibres can also be made from two different polymer compositions, e.g. PET/PA, PA/Spandex, etc. Different types of polymers, such as polyester, polyamides, polystyrene, polyurethane, polyolefines, polylactic acid, co-polyamides and soluble co-polyesters, have been used to spin bi-component fibres.

6.2 Classification of bi-component fibres

Bi-component fibres can be classified into three types:

1) **Side-by-side (S/S)**: where each polymer is divided along the length into two or more distinct regions of the cross-section.
2) **Core sheath type (C/S)**: In this type of fibre one of the components is completely surrounded by another component. Core sheath type fibres can be self-bonding (e.g. two components with varying melting points with the lower melting point polymer on the sheath), surface tailored fibres (e.g. sheath containing expensive additives) or filled fibres (e.g. a core of recycled material covered by a sheath with the desired properties).
3) **Matrix fibril (M/F)** polyblend type: where many fine fibrils of one polymer are dispersed randomly in size and location but with axial alignment in a matrix of another component.

- Fibres produced by the third type are referred by Richeson and Spruiell[1] as bi-constituent, while the term 'bi-component' is applied only to fibres of the first and second categories. There are many variants of bi-component fibres with varying phases (in terms of quantity and location) and tailored interfaces. Bi-component fibre spinning comes under the concept of blend spinning and has been well detailed by Salem.[2] Depending on the choice of constituents, they can be either compatible or non-compatible. The first category results in a homogenous single phase solution, and the spinning resembles the spinning of single component

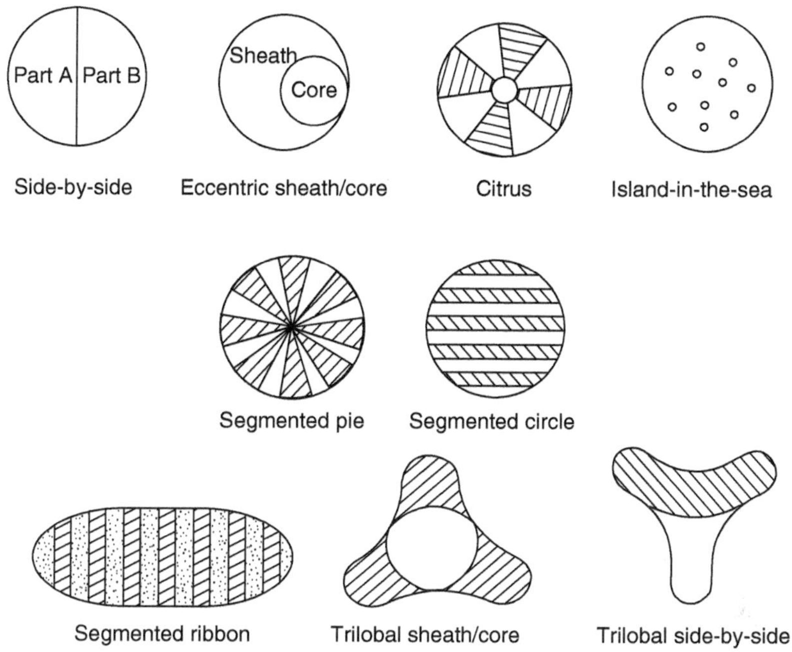

6.1 Cross-sectional conformations of different types of bi-component fibres

homopolymers. Incompatibility between the polymers results in phase separation. The control of morphology becomes more challenging, and further research findings will be subsequently discussed to demonstrate the complexity. An excellent review of cross-sections of the different bi-component fibres has been carried out by Hills Inc.[3] Structure-wise, the bi-component fibres have undergone significant development and, as an extension to the basic structures, can have interesting cross-sections (Fig. 6.1) such as:

- Citrus, wedge, or segmented-pie;
- Hollow or non-hollow;
- Regularly round cross-section; and/or
- Irregularly non-round cross-section, including flat ribbon, multilobal, triangle, paralleled strip, etc.

Every type of fibre has specialist end uses. Side-by-side fibres and eccentric core/sheath fibres possess the often desired property of self-crimping. Drawing or heating the sides of such fibres causes the fibre to wrap itself into a helical configuration, the intensity of which will be a function of the thermal behaviour of the materials used and their arrangement. The core/sheath fibres are able to retain individual properties of each component. C/S fibres with polypropylene sheath around a nylon core may potentially produce a fibre with the wear resistance of nylon and display the stain resistance of polypropylene. In the melt-blown industry, sheath-core fibres are used as a thermobonding fibre. A strong bond must be made between the two materials to prevent fibre splitting. For an interesting analysis of splitting of various fibres, including segmented ribbon, the reader can refer to an article by John Hagewood.[4] A fibre made of chemically different species will require a specialist scheme to enhance the strength of the interface. The islands-in-the-sea fibres are one of the most common methods used for high efficiency filtration. The splitting of the fibres is carried out by thermal, mechanical or chemical means. Trilobal fibres are used for speciality filtration purposes.

6.3 Spinning technologies

The first bi-component fibres were spun by Dupont in 1960; a side-by-side fibre made from two types of polyamide fibres with different retraction.[5] However, the initial manufacturing process involved complex and expensive spinning and was technically and economically unsatisfactory. Extensive research by such companies as Hills Inc. and Enca Inc. has resulted in a modified design of the spinning system using thin flat plates with grooves and holes to conduct the polymer, and thus the process became flexible and effective. The basic system consists of two feeding systems, two polymers to the spin packs and a distribution system to meter both polymers to the die. The feed systems may vary and are generally a combination of:

1. Two continuous polymerisation processes.
2. Two extruders.
3. A combination of (1) and (2).

An excellent review of developments can be found in the study carried out by Fitzgerald *et al.*[6] The earliest 'synthetic' bi-component fibres were spun from viscose by Sisson and Morehead.[7] The process involved extruding two different component solutions (or dopes) through the same spinneret hole (Fig. 6.2). The two components A and B are fed to a single spinneret hole split by a knife edge or septum which channels the two components into a side-by-side or bilateral arrangement. This same principle can be applied to multi-ring/multi-knife edge arrangements and to non-circular arrays of holes and knife edges.

Figure 6.3 illustrates a spinneret design for preparing sheath-core conjugate fibre. One of the component streams is fed through a narrow tube while the second component stream supplying the sheath is fed around this

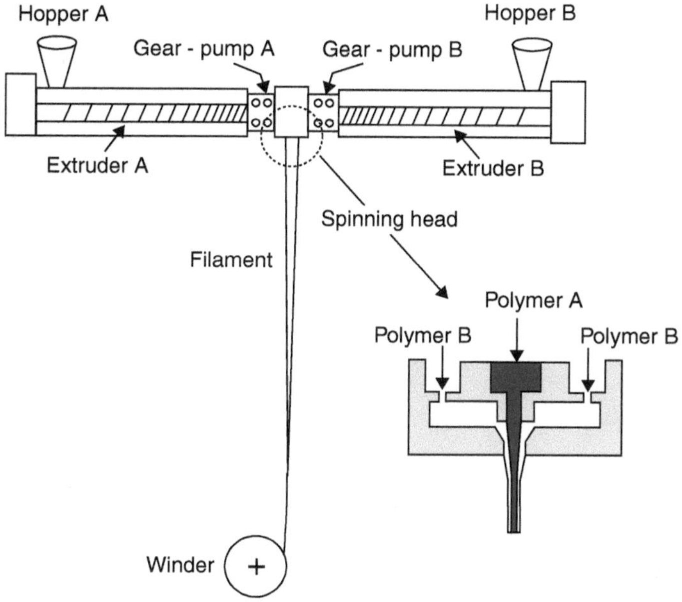

6.2 Schematic of bi-component spinning. (After Reference 4.)

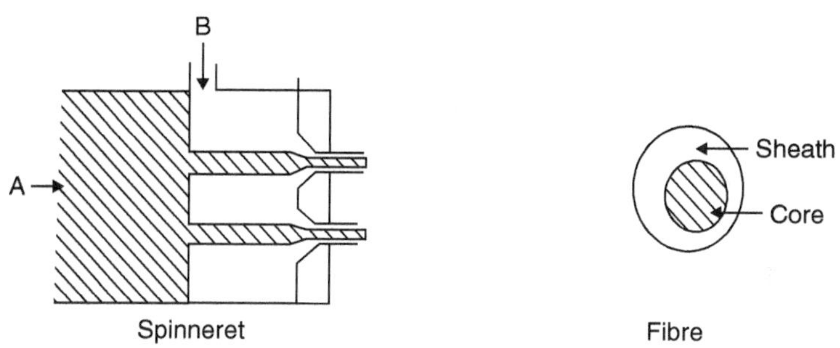

6.3 Core conjugate spinning. (After Reference 4.)

narrow tube, and the combined stream is emitted from a single hole. The combined material which emerges is coagulated in a sheath-core arrangement. The eccentricity of the sheath-core necessary for crimp development is ensured in two ways. Firstly, normal variability in coagulation conditions is sufficient to ensure crimping. Secondly, a misalignment of the centre of the spinneret hole and the tube results in eccentricity. For sheath-core spinning of bi-component fibre, separate provision has to be made for merging the component streams at each spinneret hole.

Figure 6.4 illustrates the concept of a 'pipe-in-pipe' mixing element. One of the component streams envelops the other at the end of the tube of the

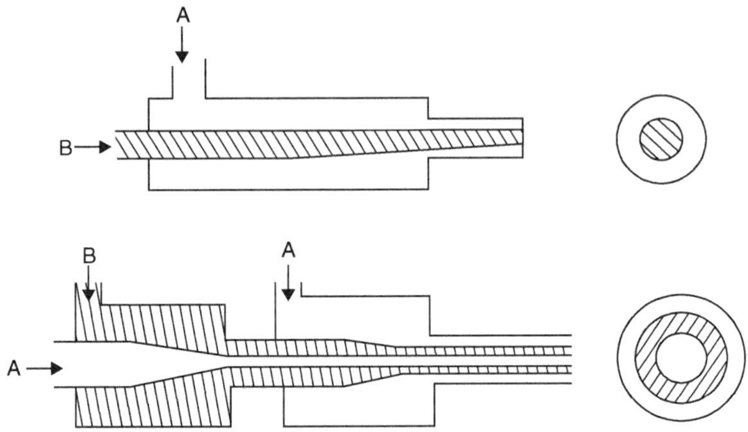

6.4 Pipe-in-pipe mixing element. (After Reference 4.)

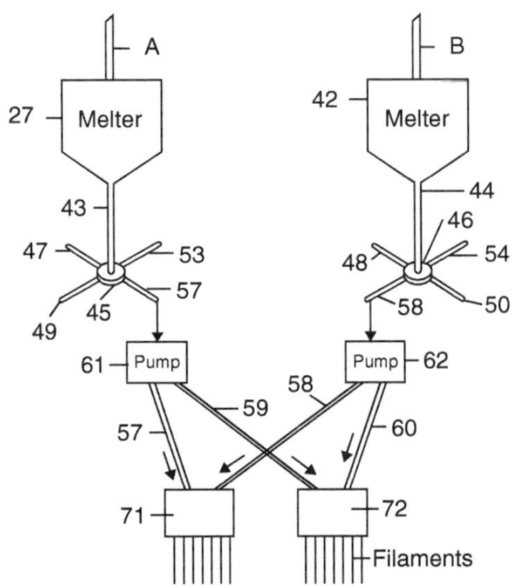

6.5 Schematic diagram of bi-component spinning. (After US Pat. 3381074.)

inner stream. The resultant stream contains in cross-section the individual streams in concentric circles. Streams can be further added by consecutive mixing elements to provide merged dope streams of concentric layers of the two components. The relative amount of material in each stream is determined by the volumetric rate at which each successive stream is pumped.

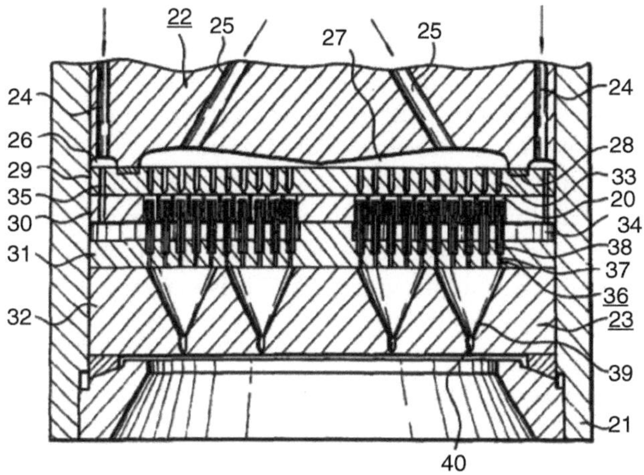

6.6 The spinneret assembly for islands-in-sea spinning. (After US Pat. No. 3692423.)

It is interesting to note the developments in the spinning designs of bi-component spinning. One of the first patents in bi-component spinning went to Okamoto from the US Pat. No. 3531368.[8] As seen in Fig. 6.5, the melters 41 and 42 are for two different polymers through metering pumps 61 and 68. Transfer pipes 57 and 59 transfer dual streams from the left hand pump to the spinneret assembly, from where the bi-component fibres are extruded.

In sheath-core type bi-component spinning, Okamoto incorporated an improvement in US Pat. No. 3692423[9] by changing the converging angle of the funnel-shaped space from θ to 75° (Fig. 6.6).

Conduit 24 supplies the sea constituent polymeric liquid, whereas the island component is supplied through conduit 25. Chamber 27 distributes the island constituent polymeric liquid into the spinneret 23. The chamber 26 is fluid tight and separated from chamber 27 by annular packing 28. The converging angle of 75° is effective for gradually decreasing the diameter and uniformly uniting the composite streams in a stabilised condition. There have been several further improvements in spinneret design. One significant development was by Moriki and Ogasawara,[10] who came up with a noteworthy improvement in the spinneret design. The main problem encountered with earlier spinnerets concerned the control of viscosity for the islands and sea components. It was found in many cases that the island component fused with the sea component. The new design allowed the constituent streams, substantially engrafted in a sea constituent, to flow in such a manner that they were uniformly distributed and separated from each other in a cross-sectional view.

6.4 Factors affecting bi-component and bi-constituent spinning

Choice of polymers is a highly crucial factor in bi-component spinning. If the viscosities of the polymers are close, a continuous phase would exist. The morphology would strongly depend on the relative amounts of each phase and the relative viscosity of the two polymers. It has been observed that when polymers have varying viscosity, the lower viscosity polymer tends to encapsulate the one with a higher viscosity. This is extremely important for sheath-core spinning where, in the final fibre, the sheath is of lower viscosity polymer and the core has a higher viscosity. Final diameter of fibrils, in case of matrix fibril type bi-constituent fibres, can be achieved through the control of particle size of the discontinuous phase, the amount of spin draw, and post spinning parameters. Some of the most important parameters, along with the relevant research, are outlined below.

6.4.1 Viscosity and structure development

Large viscosity differences between the two materials can also affect the spinnability of the fibre as it exits the spinneret. Differences in viscosity and position in the capillary lead to differing velocities for the two materials. As the two polymers traverse the capillary, they will spontaneously rearrange themselves until the configuration that produces the least amount of resistance to flow is obtained. Material of lower viscosity will move to regions of greater shear, near the walls, leading to the encapsulation effect. The properties of such fibres also depend on the structure of individual components. Kikutani *et al.*[11] have shown that the orientation and crystallinity of each component in a bi-component system can be different from the single components. The polyester core of the PP/PET sheath/core fibres has better crystallinity than the single components. With polypropylene, a lower density remains in the sheath and has a much lower orientation and a smectic structure, even at high take-up speeds. Simulation carried out by the researchers suggested that solidification stress in the polyester component increased, while the PP component decreased. This difference has been suggested as a result of the difference in activation energy of elongational viscosity of the two polymers.

In another work by Shi *et al.*[12] sheath-core type poly(butylene terephthalate)/poly(butylene adipate-co-terephthalate) (PBT/PBAT) bi-component fibres were produced by extruding the melts of PBT (intrinsic viscosity of 0.88 dL/g) as the sheath and PBAT (melt volume flow rate of 3–6 mL/10 min) as the core, using two different extrusion systems. The co-axially combined polymer melts were extruded through an annular-type spinneret at a temperature of 260°C, and the mass flow rate was controlled at 3 g/min for each component. When comparing the development of crystal structure of the individual components in PBT/PBAT bi-component spinning with that of the corresponding single component spinning, WAXD patterns of the as-spun fibres prepared at 3 km/min showed sharper diffraction spots of PBT in PBT/ PBAT fibres than the diffraction arches of PBT in single component fibres. Compared with respective single component fibres, in PBT/ PBAT fibres, the molecular orientation and crystalline structure development of the PBT component were greatly enhanced, whereas the molecular orientation of the PBAT component was greatly suppressed.

It is interesting to note that the authors had reported an improved spinnability of PBT/PBAT fibres compared with the single component spinnings of PBT and PBAT. The enhanced structure of the PBT component led to an improvement in both thermal property and tensile strength, suggesting that it is an effective way to promote the structure development of PBT through bi-component spinning with PBAT. The PBAT component was found to have nearly non-oriented structure in both crystalline and amorphous phases, which is a desirable feature for denier-decreasing treatment in fabric applications.

The authors clarified mutual interaction using numerical simulation between the two polymer melts during bi-component spinning. If two polymers (A and B) have a difference in activation energy of elongational viscosity or solidification temperature or initial elongational viscosity, then one of the polymers undergoes concentrated thinning behaviour near the spinneret compared to the other component in their respective single component spinnings. The authors further clarified that in bi-component spinning, the thinning curve of the spinning line should be in-between the thinning curves of their single component spinnings. Compared with single component spinning, the higher shearing component will experience higher stress near the solidification point in the spinline due to a larger strain rate, and thereby enhanced structure development. In the case of PBT/PBAT fibres, the PBT component underwent solidification at the point where PBAT was still in the melt state, and thereby PBAT underwent orientation relaxation and remained close to the random state. PBAT remained tacky along the spinline and made the filament difficult to be taken up when it was arranged in sheath part. Non-oriented crystal structure of PBAT was speculated to be formed off spinline as its glass transition temperature (~ –30°C) is much lower than room temperature.

6.4.2 Interfacial interaction

Research by Ahmed[13] focused on a combination of poly(butylene succinate/L-lactate) (PBSL) and poly(butylene terephthalate) (PBT) as the biodegradable and non-degradable polymers. For conjugate fibres with PET as the core, the interaction between the core and sheath was very strong because of the interfacial reaction of the polymers during melt-spinning. This strong interfacial interaction has a profound effect on the physical and mechanical properties of the conjugate fibres. During drawing and annealing of the conjugate fibres, the PET core filament gains a higher crystallinity compared with the single PET fibre, while the sheath polymers retain an amorphous state. The authors demonstrated that both alkali and enzymatic hydrolyses are effective for decreasing the denier of the polyester conjugate fibre. It was interesting that conjugate fibres had much higher strength and elastic limits than the single PET fibres, indicating that stress was homogeneously applied to the core filament in both conjugate fibres. Covering the PET core filament with the PBSL or PLLA sheath reinforces the core filament, probably because the stress concentration can be reduced by the elasticity of the sheaths. On reaching the yielding point, the stress–strain curve of the PET–PBSL fibre reveals a uniformity of stress up to 20% in strain, and from this point, the stress starts to increase linearly. This was also due to the highly elastic nature of the PBSL sheath. Stress of the PET/PLLA fibre increases steadily beyond the yielding point because of the stiffness of PLLA. The cross-sectional area of both fibres showed no significant interface dividing the core and sheath, indicating excellent adhesion between the two.

6.4.3 Compatibility of polymers for sheath-core spinning

In research conducted on the sheath-core type bi-component spinning[14] of various combinations of polymers such as poly(ethylene terephthalate) (PET)/polypropylene, PET/polyethylene, PET/polystyrene, high and low molecular weight PETs, etc., it was found that the thermal and stress histories experienced by the two polymers coextruded to form a single filament are significantly different in comparison with the corresponding single component spinning. The relative differences in the inherent polymer characteristics, such as solidification temperature, activation energy of elongational viscosity, and the initial viscosity, are found to be some of the major factors influencing the mutual interaction between the component polymers. When two polymers with the above-mentioned differences in their properties are coextruded to form a single filament, the elongational stress development, and therefore the orientation and orientation-induced crystallisation development in the component that has higher solidification temperature, higher activation energy, and higher initial viscosity, was found to be enhanced. Mechanisms for these mutual interactions were clarified by investigating the spinline dynamics of the bi-component spinning through on-line measurements and process simulations. These studies led to the conclusion that, by selecting a suitable combination of polymers with different inherent properties mentioned above, it was possible to design the spinline dynamics and therefore structure development of the individual component in high-speed bi-component spinning.

In a study by Bosak *et al.*[15] the core polymer was polypropylene and the sheath material consisted of immiscible blends of two polymers. Polystyrene (Styron 484, designated as PS) and two polyamide-6 polymers (BS-400 and BS-700, designated as PA6) were used as the dispersed particles in the blend that was used as the sheath. Polyethylene (Affinity SM 1300 PE) was used as the continuous phase in the sheath for all results reported here. We also conducted preliminary studies using PP as the matrix phase in the sheath. However, the PP matrix containing PS inclusions could not be spun into fibres at any of the drawing speeds, and could be spun at the lowest speed for the PA6 inclusions. Consequently, the researchers did not investigate PP further as the matrix material for blends used as the sheath. Fibres made of pure polypropylene were used as control specimens.

When used as the sheath component in a bi-component fibre, immiscible polymer blends produce an irregular surface. The extent of surface irregularities was affected by the microstructure of the dispersed phase, *i.e.,* fibrillar PA6 as opposed to ellipsoidal PS. The microstructure also affected the tensile properties of the bi-component fibres. The static coefficient of friction was not affected by the nature of the dispersed phase, but was affected by the sheath matrix polymer (PE vs PP). As fibre surface irregularities increased, mechanical properties decreased. Bi-component fibres with PS had greatly increased surface irregularities compared to the control fibres, but also displayed lower strength and strain-to-failure. In contrast, bi-component fibres containing PA6 dispersed phase had slightly decreased strength but significantly higher strain-to-failure, but these fibres provide only a minor increase in surface roughness. The study illustrated the range and combination of surface properties that can be generated by using incompatible blends as the sheath material in bi-component fibres.

6.4.4 Post spinning conditions

In a work by Tae *et al.*[16] the effect of process conditions on crimp contraction in the side-by-side bi-component spinning and drawing of PET and modified PET was studied. Draw ratio, heat-set temperature and drawing temperature were the main effective process variables. The difference in shrinkage between two polymers was found to be the main cause of crimp contraction. Difference in elongation affected crimp contraction in the effect of draw ratio. Although the authors reported no dimensional change with heat-set temperature and drawing temperature, they had an influence on the crimp contractions. Maximum crimp contraction was 42% at a draw ratio of 3.1, heat-set temperature of 125°C, and drawing temperature of 80°C.

In another work by Tae *et al.*[17] the effects of process conditions on crimp contraction in the side-by-side bi-component spinning and drawing of two PTTs of different viscosities were studied. PTTs with IVs of 1.02 and 0.92 were selected to make a latent crimp yarn. The draw ratio, heat-set temperature, and portion of PTT-H were the main effective process variables. An analysis of the effect of spinning and drawing conditions on the crimp contraction showed that draw ratio was the most critical variable in controlling it. Increasing the draw ratio caused a difference in the shrinkage between the two parts of PTT and caused self-crimping of the fibres. The obtainable maximum crimp contraction was about 50% at a draw ratio of 1.4 and a heat-set temperature of 200°C. Although there was no dimensional change with the heat-set temperature and portion of PTT-H, they had an influence on the crimp contraction. As the heat-set temperature and portion of PTT-H increased, the crimp contraction of the latent crimp yarn also increased.

6.5 Applications of bi-component fibres

The advantage of bi-component fibres lies in their enhanced functionality. The fundamental sheath/core cross-section is useful in many applications demanding engineering polymers. In such structures, the fibre's core can be prepared with a suitable low cost polymer to deliver all of the benefit of the more expensive polymer at a material cost well below that of a fibre made from an expensive surface polymer alone. Side-by-side bi-component fibres have self-crimping properties[18] and rely on the difference in shrinkage between the two polymers. During the fabric formation process, if the fibres are not physically constrained, shrinkage can be induced by the application of heat. Since the two polymers shrink at different rates, the material behaves like a bimetallic strip and resultant shrinkage occurs.[19] Freudenberg's Evolon fabric uses segmented-pie or side-by-side bi-component fibres, which are formed and then fed directly to a hydroentangling unit where bonding and splitting occur simultaneously. The smallest diameter fibres, achieved through the sea-island process of bi-component spinning, are smaller in diameter when compared to the split spinning process. Different cross-sections, and various bi-component geometries including eccentric sheath core, were some of them, and even using polymers of varying molecular weight[20] and different forms of the same polymer. In some of the bi-component fibres, elastomeric material is used to achieve the self-crimping properties. Fibres manufactured from two different melting points are used for point-bonded non-wovens. Based on structure, the applications have been detailed by Jeffrey Dugan.[21]

The applications of bi-component fibres are many, including:

1. Increased dyeability
2. Decreased flammability
3. Improved hand and lustre
4. Improved water absorption
5. Increased light and heat stability
6. Manufacture of microfibres
7. Fibres for ion exchange and separation applications
8. Improved strength and elastic recovery.

With new research, new application areas have emerged. In a study by Shim *et al.*,[22] a number of photoluminescent sheath/core bi-component fibres were produced by an interesting process. Fine phosphor particles comprising an alkaline earth metal aluminate activated by europium with a free-flow agent were mixed and compounded to produce a polymer pellet with concentrated phosphor. The mixture of the phosphorescent polymer pellet and the same polymer without phosphorescent pigment was extruded as the core component. The concentration of photoluminescent pigment in the core was achieved by varying the ratio of the prepared phosphorescent and non-phosphorescent polymer pellet. Resultant bi-component fibres had a photoluminescent core with varied pigment concentration and a non-photoluminescent sheath as no phosphorescent polymer pellet was used in the sheath component. It was observed that the presence of photoluminescent pigment changes the tone of the core component, making it appear darker than the sheath. Distinction between the core and the sheath was not obvious with fibres having a low concentration of photoluminescent pigment in the core component. At 5% core pigment concentration, boundaries of the core/sheath were hardly seen, since the sheath and the core polymers were identical. The only difference, it was noted, was the presence of black dots representing concentrated photoluminescent pigments in the cross-section of the core. At 10% core pigment concentration, the core was observed to be slightly darker than the sheath. In nylon and PET fibres with 30% core pigment concentration, a distinct boundary between the sheath and the core layers was revealed, and the core component looks much darker since it contains a high concentration of photoluminescent pigment. The tenacity and Young's modulus of the individual fibres showed that higher loading of the photoluminescent pigments in the core component did not affect the tensile properties of the samples.

6.6 Future trends

Bi-component melt-spinning has been widely applied to produce functional and novel fibres such as hollow fibres, electrically conductive fibres, micro/nanofibres and thermal bonding fibres, to name a few. Research on biodegradable polymers has resulted in fabrics woven by sheath-core type bi-component fibres being composed of PBT and biodegradable polymers such as poly (L-lactide), poly(butylene succinate) and poly(butylene succinate-co-L-lactide), etc. The main merit of such fabrics is that, with the presence of biodegradable components, denier-decreasing treatment can be conducted in an environment friendly way. Accordingly, PBAT/PBT fibres are also expected to be applied in these woven fabrics, as they are not only environment friendly but also stretchable.

Recently, tricomponent spinning systems have been developed to coextrude three different polymers into each fibre. Interestingly, some bi-component cross-sections have been utilised in spunbond fabrics, in which filaments are extruded directly into a non-woven web without forming fibres as an intermediate product. The precision of polymer control to form the cross-section will continue to advance. After persistent research, Hills has been able to produce spin packs capable of stuffing hundreds of islands into each fibre cross-section, which enables the production of submicron microfibres. The future thus lies in further reduction of the ultimate denier of the fibres, improved spinning processes with more control, and exploration of more interesting applications for bi-component fibres with varying polymers and processes.

6.7 References

1. G. C. Richeson and J. E. Spruiell (1990), Preparation, structure, and properties of copolyester-ether elastic filaments, *J. Appl. Polym. Sci.*, **41**, 845.
2. D. R. Salem (2001), *Structure Formation in Polymeric Fibers*, Carl Hanser Verlag, Munchen.
3. http://www.hillsinc.net/bicointro.shtml (accessed 9 February 2012).
4. http://www.hillsinc.net/splitting.shtml (accessed 9 February 2012).
5. E. A. Tippets (1967), Fiber engineering to meet end use requirement, *Textile Res. J.*, **37**, 524–533
6. W. E. Fitzgerald and J. P. Knudsen (June 1967), Mixed-stream spinning of bicomponent fibers, *Textile Res. J.*, **37**(6), 447–453, Conventional Bicomponent Spinning.
7. W. A. Sisson and F. F. Morehead (1953), The skin effect in crimped rayon, *Textile Res. J.*, **23**, 152–157.
8. M. Okamoto, K. Watanabe, Y. Nukushina and T. Aizawa (1970), Synthetic filaments and the like, US Pat. No. 3,531,368 (to Toyo Rayon Inc), 29 September 1970.

9. M. Okamoto, K. Ashida, K. Watanabe and S. Taniguchi (1972), Apparatus for spinning synthetic islands-in-a-sea type composite filaments, US Pat. No. 3,692,423 (to Toray Industries, Inc., Tokyo, Japan), 19 September 1972.
10. Y. O. Moriki and M. Ogasawara (1984), Spinneret for production of composite filaments, US Pat. No. 4,445,833 (to Toray Industries, Inc.,Tokyo, Japan), 1 May 1984.
11. T. Kitukani, J. Radhakrishnan, S. Arikawa, A. Takaku, N. Okui, X. Jin, F. Niwa and Y. Kudo (1996), High-speed melt spinning of bicomponent fibers: Mechanism of fiber structure development in poly(ethylene terephthalate)/polypropylene system, *J. Appl. Polym. Sci.*, **62**, 1913–1924.
12. X. Q. Shi, H. Ito and T. Kikutani (2006), Structure development and properties of high-speed melt spun poly (butylene terephthalate/ poly (butylene adipate-co-terephthalate) bicomponent fibers, *Polymer*, **47**, 611–616.
13. A. El-Salmawy and Y. Kimura (2001), Structure and properties of bicomponent core-sheath fibers from poly (ethylene terephthalate) and biodegradable aliphatic polyesters, *Textile Res. J.*, **71**, 145.
14. J. Radhakrishnan, H. Ito, T. Kikutani and N. Okui (January 1999), Enhancement of fiber structure formation of a liquid crystalline copolyester via ultra-high speed bicomponent spinning with poly (ethylene terephthalate), *Polym. Eng. Sci.*, **39**(1), 89–98.
15. D. R. Bosak, A. A. Ogale and J. Van Dun (2005), Bicomponent fibers derived from immiscible polymer blends, *Textile Res. J.*, **75**, 50.
16. T. H. Oh (2006a), Melt spinning and drawing process of PET side-by-side bicomponent fibers, *J. Appl. Polym. Sci.*, **101**, 1362–1367.
17. T. H. Oh (2006b), Effects of spinning and drawing conditions on the crimp correction of side-by side poly (trimethylene terephthalate) bicomponent fiber, *J. Appl. Polym. Sci.*, **102**, 1322–1327.
18. M. Okamoto (1973), Spontaneously crimping synthetic composite filament and process of manufacturing the same, US Patent 3718534, to Toray Industries, Inc., Tokyo, Japan, 27 February 1973.
19. http://www.textileworld.com/Articles/2010/July/July_August_issue/Features/Specialty_Markets_Bicomponent_Fibers.html (accessed 9 December 2011).
20. V. Nadkarni (2004), Self-crimping multicomponent polymer fibers and corresponding methods of manufacture, US Patent 6811873, Hills Incorporated, 2 Nov 2004.
21. R. D. Harris, J. P. Parker and C. L. King (1988), BASF Corporation, US Patent 4732809, 22 Mar 1988.
22. E. Shim, B. Pourdeyhimi and T. J. Little (2004), Luminescence and mechanical properties of photoluminescent core bicomponent fibers, *Textile Res. J.*, **74**, 982, Materials and fiber cross sections.

7
Electrospinning, processing and characterization of polymer-based nano-composite fibers

Q. Q. NI, X. D. JIN, H. XIA and F. LIU,
Shinshu University, Japan

DOI: 10.1533/9780857099174.2.128

Abstract: Electrospinning is a relatively simple and inexpensive method to produce fibers with diameters in the nanometer range. In recent years, many different kinds of polymer nanofibers have been fabricated by electrospinning technology, including nano composite fibers. The challenges of developing multifunctional nano-composite fibers are reviewed in this chapter. $CoFe_2O_4$ nanoparticles with a particle size of about 10–25 nm were synthesized using a simple hydrothermal process. $CoFe_2O_4$/PAN (polyacrylonitrile) nanofibers with various $CoFe_2O_4$ loadings were prepared by electrospinning, and the fiber diameter was controlled to a distribution of between 200 and 400 nm. Then, PAN nano-composite fibers containing carbon nanotubes (CNTs) and $CoFe_2O_4$ nanoparticles were fabricated by electrospinning. The presence of $CoFe_2O_4$ nanoparticles and CNTs in the PAN matrix was confirmed by X-ray diffraction (XRD) patterns, Raman spectra, SEM, and TEM observation. As a multifunctional material, the nano-composite fibers achieved both EMI SE and magnetic performance for the developed PAN/CNT-$CoFe_2O_4$ fibers. This performance can be accurately tailored by adjusting $CoFe_2O_4$ nanoparticle loading.

Key words: electrospinning, magnetic nanoparticle, carbon nanotube, PAN, EMI SE.

7.1 Introduction

Electrospinning is a relatively simple and inexpensive method to produce fibers with diameters in the nanometer range.[1–4] In recent years, many different kinds of polymer nanofibers have been fabricated by electrospinning technology, including nano-composite fibers. Nano-composite fibers need to be multifunctional with engineering applications in such areas as sensing, electroactive actuators, etc. The challenges of developing multifunctional nano-composite fibers are reviewed in this chapter.[5]

7.2 Principles of electrospinning

In an electrospinning system, a high electrostatic voltage is passed through a drop of polymer solution held by its surface tension at the end of a capillary, as shown in Fig. 7.1. The surface of the liquid is distorted into a conical shape, known as the Taylor cone. Once the voltage exceeds a critical value, the electrostatic force overcomes the solution surface tension, and a stable liquid jet is ejected from the cone tip. The solvent evaporates as the jet travels through the air, leaving behind ultrafine polymeric fibers that collect on an electrically grounded target. The jet often follows a bending or a spiral track, resulting from the interaction between the external electric field and the surface charge of the jet. The bending instability of jets not only results in the electrospinning jet being elongated up to ultrafine fibers, but also leads to the formation of randomly deposited nonwoven electrospun fiber mats.[6–8]

Electrospun mats have a larger specific surface area and smaller pore size compared to commercial nonwoven fabrics. They are of interest in a wide variety of applications, including semi-permeable membranes, nanocomposites, filters, protective clothing and biomedical applications such as wound dressings, tissue engineering scaffolds, and drug delivery systems.[9–12] Recently, unidirectional-like nanofiber mats fabricated by control collector speed were reported and a high mechanical property was obtained.[13]

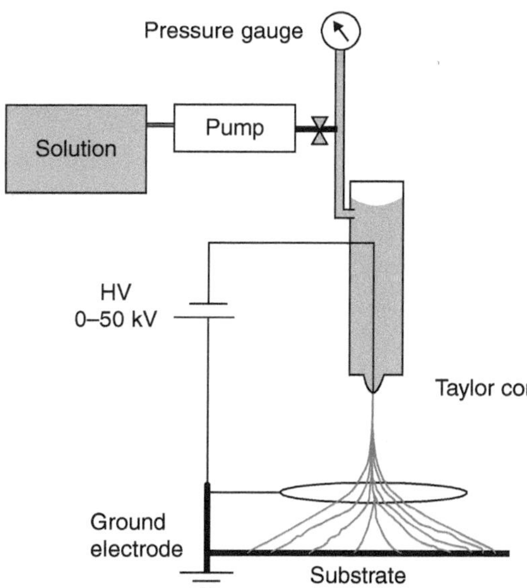

7.1 Set-up for electrospinning from polymer solutions.

In electrospinning technology, the control of fiber diameter is the key. The diameter of fibers formed during electrospinning is influenced by:

- System parameters:
  - Polymer properties, such as the molecular weight, structure and polydispersity of the polymer, and melting and glass transition points.
  - Solution properties, such as solvency, volatility, viscosity, conductivity, surface tension, and the presence of other additives.
- Process parameters:
  - Ambient parameters, such as solution temperature, humidity, atmosphere, and air velocity in the electrospinning chamber.
  - Equipment parameters including voltage, field strength, electrode distance and arrangement, flow rate, delivery volume, and needle diameter.

The formation of fibers in the electrospinning process is also influenced by surface tension derived from the applied electrical charge to the polymer.

A further advantage of electrospinning, compared to conventional solvent spinning, is that water can be used as a solvent. Water-soluble fibers have to be cross-linked, e.g. by thermal or by chemical cross-linking.[14,15,16] These advantages result in the application potential of nanofibers in broad fields such as separation, adsorption, filtration, catalysis, fiber reinforced composites, tissue engineering, wound dressings, drug delivery systems, sensors, cleaning tissues, and protective textiles among others.[14,16–20,22–24] Nanofibers can be spun from polymer solutions or from polymer melts. Larrondo and Manley[25–27] were the first to carry out and report on melt electrospinning experiments. Working with polyethylene (PE) and polypropylene (PP) in the early 1980s, they successfully formed fibers with diameters in the tens of microns range. Electrospinning from polymer melts has the advantage of not requiring the solvents which, in other processes, would have to be removed by evaporation. However, the melting temperature of the polymers is an important factor in determining whether the procedure can be used to produce nanofibers.

7.3 Electrospinning technology

A typical electrospinning equipment has three components: a high voltage source, a spinneret (or nozzle), and a collector (as shown in Fig. 7.2). The polymer solution or melt is applied in a syringe equipped with a piston and a stainless steel capillary serving as electrode, and pushed through by a pump with a defined flow rate. The spinneret is connected to a high voltage source that applies a high voltage to the polymer. This results in the formation of a polymer drop at the end of the spinneret. When a higher voltage

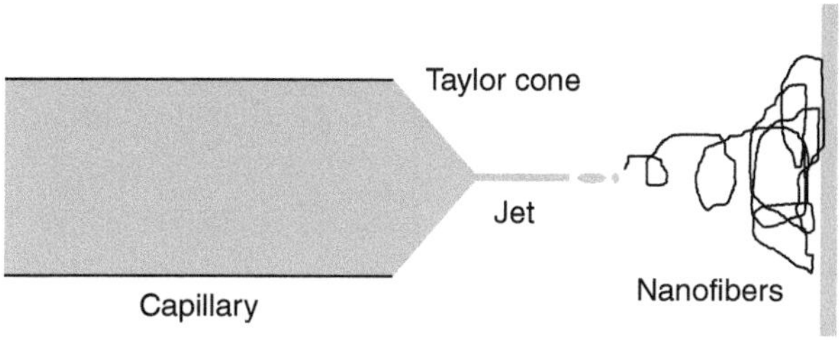

7.2 Taylor cone in electrospinning process.

is applied, the drop changes its shape and turns into a conical form (Taylor cone) (Figs 7.1 and 7.2).[25–27,28,29] At a defined voltage, the surface tension of the polymer cone at the tip of the spinneret starts to elongate and stretch, so that a charged jet is formed. The jet moves in loops, bending and whipping toward the electrode with opposite polarity or to the grounded target (Fig. 7.2).

Recent experiments have demonstrated that the rapidly whipping fluid jet is an essential mechanism of electrospinning.[30,31] Different collection systems are known.[32] Typically, in producing nonwoven mats, metal plates are used in counter electrodes and in collection systems for the nanofibers or nano-webs. In some special applications, additional grounded collectors are necessary.

Electrospinning can be used to produce novel fibers with diameters between 100 nm and 10 μm. As the solvent evaporates, the polymer molecules can come together, either by a phase separation through a spinodal reaction or through classic nucleation and growth of the crystalline phase.[33] As a result, the structure in the polymer deposited on the collector may form a totally amorphous or a semicrystalline spherulitic or textured fibrillar structure. In electrospinning, jets are stretched along their axes by the external electrical field and are elongated further by the repulsive force between charges on adjacent segments. The resulting area reduction rate and associated high longitudinal strain rate imply that the macromolecules in the fibers should be stretched and axially oriented.[34] It is generally recognized that electrospinning may lower the degree of crystallinity in the polymer.[35–37] The exact reasons for this behavior are not clear. It has been suggested that the development of structure in electrospinning occurs much more rapidly than other processes, and this kinetic effect may result in lower crystallinity.[37] A high degree of orientation may also be observed in the fibrils. The degree of orientation of the molecules in the amorphous regions is directly proportional to the amount of extensional flow. In summary, electrospinning is a novel technique that can be used to produce nano-scale porous structures with a variety of morphologies. The fiber size and distribution, and the inter-fiber spacing (i.e. porosity) and distribution, can be varied significantly by controlling the process parameters. Drugs and growth factors can be incorporated easily into the structure for biomedical applications.

7.4 Electrospinning of nanofibers

One-dimensional (1-D) nano-composite fibers (i.e. long fibers with a nano-scale diameter) have attracted considerable attention, owing to their potential applications in sensors,[38–40] as antibacterial materials,[41,42] in gas separation,[43–45] in lithium-ion batteries,[46,47] and as photocatalysts.[48,49] Electrospinning technology has proved to be a simple, efficient, and versatile way to fabricate nano-composite fibers.[50]

A number of novel electrospun nano-composite fibers have been researched and developed. Chen *et al.*[51] describe polyimide nanofiber membranes containing multi-walled carbon nanotubes (CNT) with high strength and toughness that were fabricated using electrospinning. Strongly photoluminescent electrospun nano-composite fibers combining polyvinylpyrrolidone and rare-earth complexes were prepared by Zhang *et al.*[52] In addition, several types of magnetic polymer fibers were fabricated by incorporating magnetic nanoparticles such as Fe_3O_4 and FePt into a polymer matrix and then electrospinning.[53–55] The incorporation of a component – such as Fe_3O_4 or CNT – could clearly introduce new functionality to a nano-composite fiber. Usually, however, nanofibers are limited to a single component, and there is comparatively little published material on multi-component nano-composite fibers. Thus, combinations of polymer, particles, and nanofibers may result in novel multifunctional nano-composite fibers, with a wide range of applications.

CNTs are a prospective novel material for multi-component nano-composite fibers, as they have been widely applied as composites components because of their excellent optical, mechanical, and electrical properties.[56–58] Cobalt ferrite ($CoFe_2O_4$) nanoparticles are an important magnetic material used in high-density data storage, ferrofluids, and electronic devices.[59,60]

This chapter outlines the fabrication, using electrospinning technology, of nano-composite fibers made from polyacrylonitrile (PAN) containing two other components, multi-walled CNTs and $CoFe_2O_4$ nanoparticles (PAN/CNT-$CoFe_2O_4$), and an examination of their characteristics. The incorporation of CNT and $CoFe_2O_4$ nanoparticles was expected to give rise to fibers with dual functionality, namely electromagnetic shielding and magnetic properties.

PAN, with an average molecular weight of 150 000, was purchased from Aldrich Chemical Company, USA. Multi-walled CNTs with diameter of 20–30 nm were purchased from Wako Pure Chemical Industries, Ltd., Japan. All other reagents in this work were of analytical grade and used as received without further purification.

7.4.1 Fabrication of $CoFe_2O_4$/PAN nanofibers by electrospinning

Pristine CNTs (30 mg) were first dispersed into 60 mL nitric acid (10 M) by ultrasonication for 1 h. This CNT suspension was then transferred into a round-bottomed flask equipped with a magnetic stirring bar and a reflux condenser, and refluxed at 130°C for 12 h with vigorous stirring. The treated CNTs were washed repeatedly with deionized water and ethanol to remove any residual nitric acid. The CNTs were collected by filtration and dried in a vacuum drying oven.

$CoFe_2O_4$ nanoparticles were synthesized by a hydrothermal method. First, 0.6 mmol $FeCl_3{\cdot}6H_2O$ and 0.3 mmol $CoCl_2{\cdot}6H_2O$ were dissolved into 50 mL deionized water. To this solution, sodium hydroxide was added drop-wise with vigorous stirring to adjust the pH value to about 13.5. The resulting suspension was then transferred to a PTFE-lined stainless steel autoclave and maintained at 160°C for 8 h.

The synthesized $CoFe_2O_4$ particles were dispersed ultrasonically in dimethyl formamide (DMF) for 1 h, and PAN was added into this solution. The PAN concentration was kept at 8 wt.%. After stirring, the solution was used to spin the nanofibers by electrospinning. As a comparison, nanofibers with several $CoFe_2O_4$ loadings were prepared.

7.4.2 Fabrication of PAN fibers containing CNT and $CoFe_2O_4$

CNTs and $CoFe_2O_4$ nanoparticles were prepared as described in Section 7.4.1. The treated CNTs and the synthesized $CoFe_2O_4$ nanoparticles were added to DMF, and dispersed by ultrasonication for 1 h. PAN was then dissolved into this suspension with magnetic stirring. After that, a more vigorous agitation was carried out to make sure the electrospinning solution was completely mixed. The mass fractions of CNT and $CoFe_2O_4$ nanoparticles to dissolved PAN were set at CNT: PAN = 5%, 10% and $CoFe_2O_4$:PAN = 5%, 10%. The concentration of PAN to DMF (PAN:DMF) was adjusted in the range 7–8%.

The prepared electrospinning solution was added into a 12 mL syringe equipped with a 0.80 mm stainless steel spinneret. A voltage of 8–12 kV was then applied to this solution at ambient temperature, a polymer jet was formed under the influence of the external electric field, and the fabricated nano-composite fibers were collected using a rotating disk collector. The fiber samples will be referred to by their mass fractions (CNT:PAN and $CoFe_2O_4$:PAN). The PAN/CNT-$CoFe_2O_4$ fibers are thus PAN/5CNT-10$CoFe_2O_4$ (CNT:PAN = 5%, $CoFe_2O_4$:PAN = 10%) and PAN/10CNT-5$CoFe_2O_4$ (CNT:PAN = 10%, $CoFe_2O_4$:PAN = 5%). As a control, single component PAN fibers (PAN/5CNT, PAN/10$CoFe_2O_4$) were also prepared.

7.4.3 Characterization techniques

Raman spectra were recorded using a HoloLab series 5000 Raman spectroscope (Kaiser Optical Systems, Inc., USA) at a 514 nm laser excitation. Morphologies of as-prepared $CoFe_2O_4$ nanoparticles and the electrospun fibers were observed by field emission scanning electron microscope (FESEM, S-5000, Hitachi, Japan) and scanning electron microscope (SEM, S-3400 N, Hitachi, Japan) with an operating voltage of 20.0 kV. Analyses of XRD patterns were carried out on a Rigaku Geigerflex 2028 diffractometer (Rigaku, Japan) with Cu Kα radiation (1.5406 Å). Transmission electron microscopy (TEM) observations were performed with a JEM-2010 electron microscope at 200 kV accelerating voltage. The electrospun fibers were directly collected on a copper grid with carbon film during the electrospinning process. Electromagnetic interference shielding effectiveness (EMI SE) was estimated over a frequency range of 8–12 GHz (X band) on an SE measurement system. Magnetic measurements were performed on a vibrating sample magnetometer (VSM) with a maximum magnetic field of ±10 kOe at room temperature.

7.5 Properties of electrospun nanofibers

The properties of developed electrospun nanofibers are key issues for their applications in industry. Here, the structure and morphology of the nanofibers were characterized by field emission scanning electron microscopy, transmission electron microscopy, X-ray powder diffraction, and their electromagnetic interference shielding effectiveness and magnetic property were also evaluated for electromagnetic shielding applications.

7.5.1 $CoFe_2O_4$/PAN nanofibers

The $CoFe_2O_4$/PAN nanofibers were fabricated prior to the fabrication of PAN fibers containing carbon nanotubes and cobalt ferrite[38]. The FESEM image of synthesized $CoFe_2O_4$ particles is shown in Fig. 7.3. Particles with a diameter of 10–25 nm are observed, with their irregular shapes attributable to aggregation and surfactant-free conditions.

Figure 7.4 shows the microscope images of the $CoFe_2O_4$/PAN nanofibers with various $CoFe_2O_4$ loadings (1 wt.%, 4 wt.%, 7 wt.%, 10 wt.%). The concentration of PAN to DMF (PAN:DMF) was adjusted between 7% and 8%, which is an optimum value for the nanofiber fabrication; at other values,

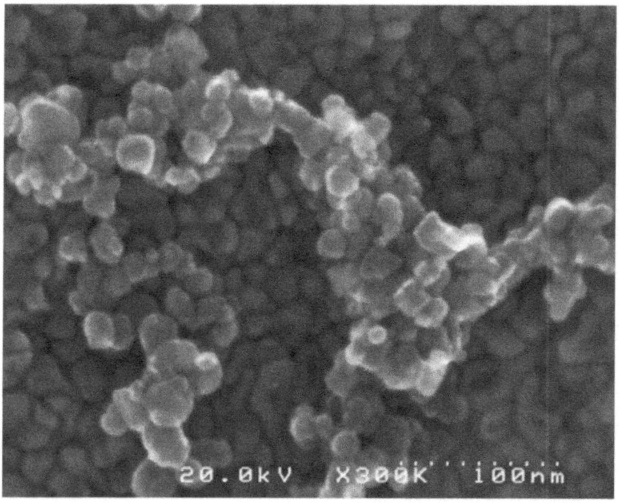

7.3 FESEM image of $CoFe_2O_4$ particles synthesized by hydrothermal method.

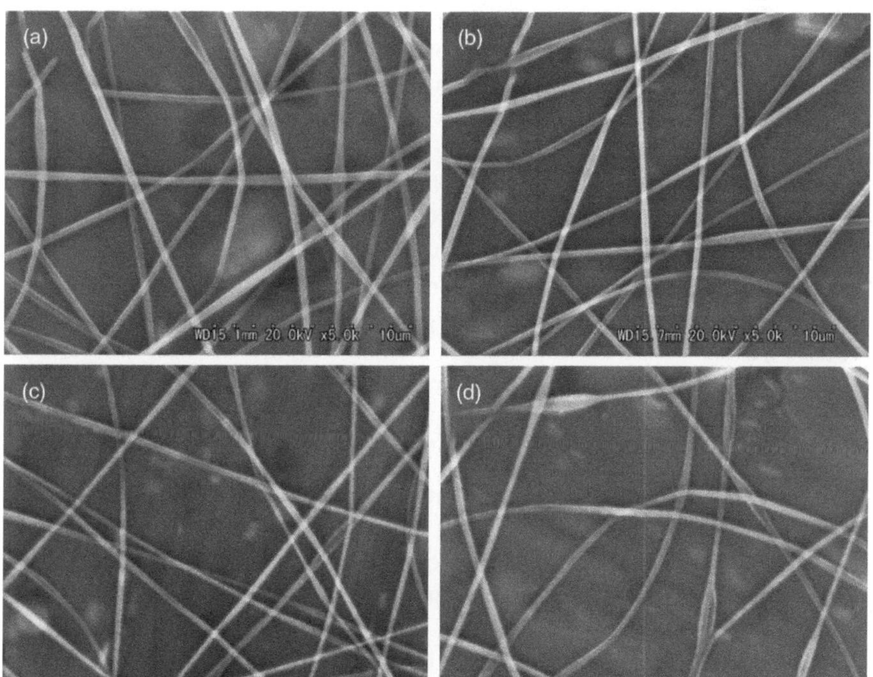

7.4 SEM images of $CoFe_2O_4$/PAN nanofibers with various $CoFe_2O_4$ loadings (a) 1 wt.%; (b) 4 wt.%; (c) 7 wt.%; (d) 10 wt.%.

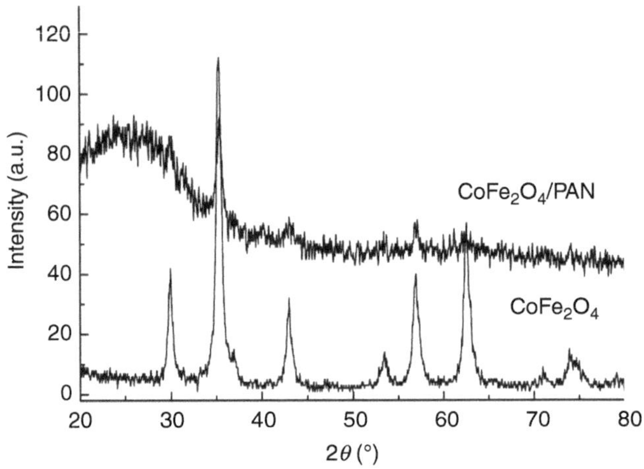

7.5 XRD patterns of $CoFe_2O_4$ and $CoFe_2O_4$/PAN nanofibers.

beading will be observable. Notably, beads begin to appear at $CoFe_2O_4$ loadings of 10 wt.%.

During $CoFe_2O_4$/PAN nanofiber preparation, the diameter distribution was controlled at 200–400 nm for all of the $CoFe_2O_4$ loadings.

As shown in Fig. 7.5, obvious peaks could be observed in the XRD pattern of $CoFe_2O_4$, which is consistent with the standard pattern of $CoFe_2O_4$. $CoFe_2O_4$ particles were successfully synthesized by the hydrothermal method. The prepared $CoFe_2O_4$/PAN nanofibers with 10 wt.% $CoFe_2O_4$ loading were also characterized by XRD; due to the effect of PAN, the intensity of the peaks observed decreased significantly. Although some weak peaks were hidden, the main characteristic peaks could be easily identified, matching the pattern of $CoFe_2O_4$, suggesting that $CoFe_2O_4$ was embedded in the PAN.

7.5.2 PAN fibers containing CNT and $CoFe_2O_4$

A uniform dispersion of CNTs in the electrospinning solution is critical for fabricating nano-composite fibers. Poorly dispersed CNTs always yield beaded morphologies, even at low concentrations.[61] However, van der Waals forces cause these CNTs to readily aggregate, making dispersion in liquid difficult.[62,63] To overcome this problem, the CNTs were treated with concentrated nitric acid, which has proven to be an effective method for improving the dispersion of CNTs in liquid.[63,64]

Raman spectroscopy was used to investigate the structures of the CNTs before and after treatment, and the spectra are shown in Fig. 7.6. For both pristine and treated CNTs, the patterns display two obvious bands centered

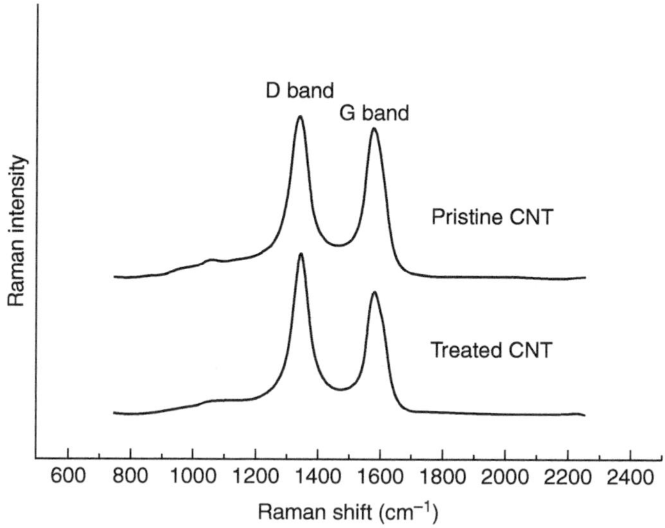

7.6 Raman spectra of pristine and treated CNTs.

around 1343 cm^{-1} and 1580 cm^{-1}, which may correspond to the D and G bands. The D band originates from disordered structures or defects in graphite carbon, and the G band could be associated with tangential stretching vibrations of the C–C bonds present in graphite.[65] Treatment by concentrated nitric acid may introduce functional groups (such as carboxyl and hydroxyl) onto the CNTs, the presence of which leads to an increase in disordered structures. Therefore, the intensity ratio of D and G band ($R = I_D/I_G$) is an index for disorder density,[66] and was used to evaluate the effectiveness of the treatment on the CNTs. From Fig. 7.6, the R-value is 1.03 for pristine CNTs, while the treated CNTs show a value of 1.32. The increased R-value indicates the presence of a more disordered structure in the CNTs after treatment, presumably owing to the introduction of functional groups. We assume that the treatment was limited to the CNT surface because of the similar Raman spectra patterns for pristine and treated CNTs shown in Fig. 7.6.

Photographs of several kinds of electrospun fiber sheets (PAN, PAN/5CNT, PAN/10$CoFe_2O_4$, PAN/5CNT-10$CoFe_2O_4$) are shown in Fig. 7.7. The photographs provide visual evidence that regular sheets of nano-composite fibers were fabricated by electrospinning. Compared with the white PAN fiber sheet (Fig. 7.7a), the PAN/5CNT and PAN/10$CoFe_2O_4$ fiber sheets (Figs 7.7b and 7.7c) are deep black and light brown in color, while the PAN/5CNT-10$CoFe_2O_4$ fiber sheet (Fig. 7.7d) shows a mixture of colors from Figs 7.7b and 7.7c. These variations in the color of the electrospun fiber sheets could be attributed to the incorporation of CNTs and $CoFe_2O_4$ nanoparticles into

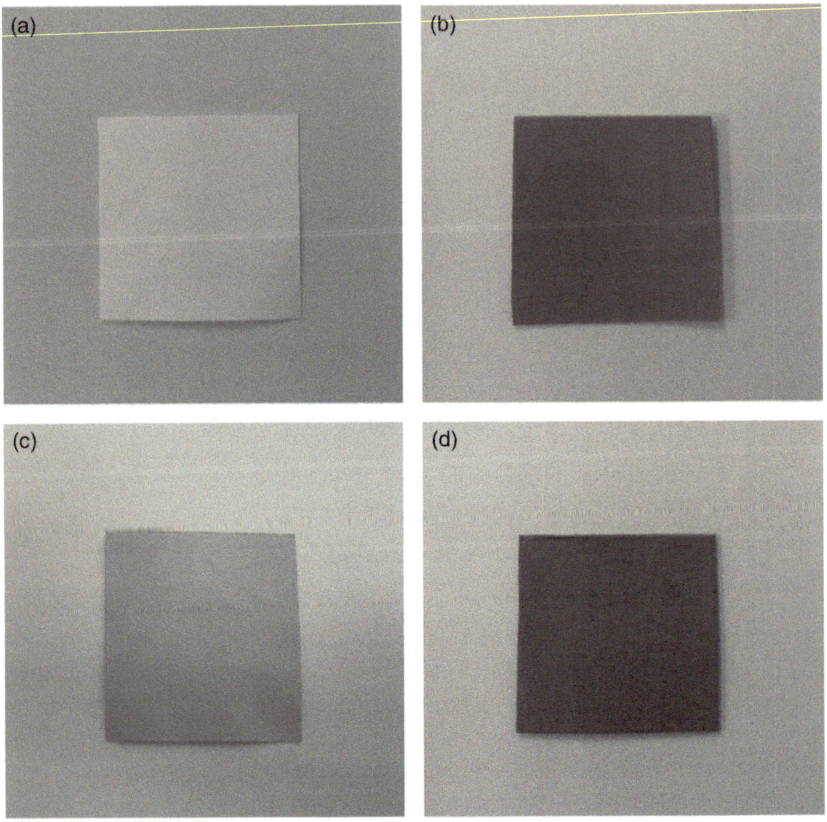

7.7 Photographs of electrospun fiber sheets (a) PAN; (b) PAN/5CNT; (c) PAN/10$CoFe_2O_4$; (d) PAN/5CNT-10$CoFe_2O_4$.

PAN. The presence of CNTs and $CoFe_2O_4$ nanoparticles in the PAN fibers are indicated in color.

The SEM images (Fig. 7.8) give further detailed information about the electrospun fibers. PAN fibers of diameter 200–400 nm display a smooth surface, and no beads are observed (Fig. 7.8a). Compared with PAN fibers, there is no obvious difference for PAN/10$CoFe_2O_4$ fibers (Fig. 7.8b), except for a slight increase in diameter. This suggests that adding $CoFe_2O_4$ nanoparticles to a PAN electrospinning solution hardly affects the appearance and size of the electrospun fibers. Figure 7.8c shows PAN/5CNT fibers with a further increased diameter of about 400–700 nm. This could be attributed to the increase of solution viscosity because of the introduction of CNTs.[67] The regular morphology of the fibers also suggests a uniform CNT dispersion through the PAN solution, even at high CNT concentrations. Fig. 7.8d illustrates PAN/5CNT-10$CoFe_2O_4$ fibers fabricated by electrospinning;

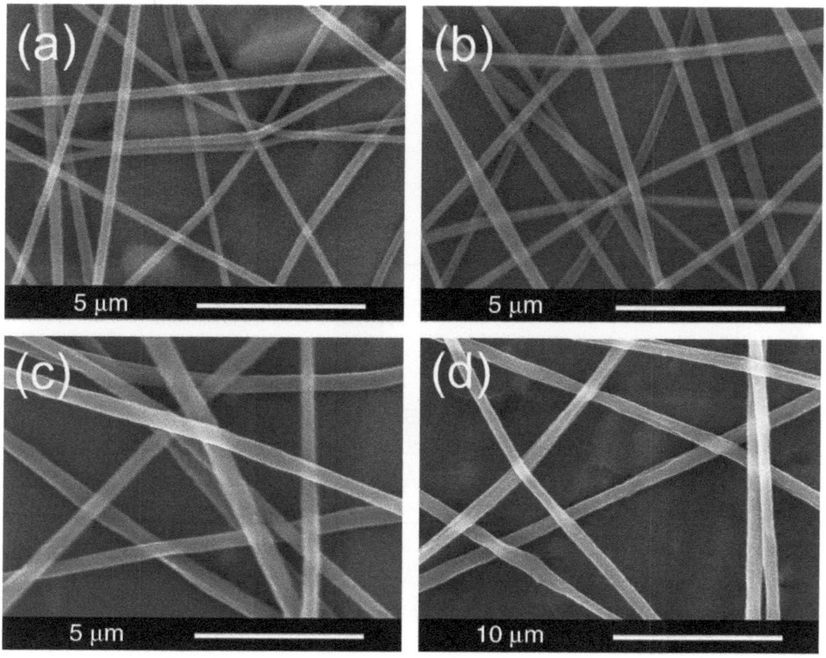

7.8 SEM images of electrospun fibers (a) PAN; (b) PAN/10$CoFe_2O_4$; (c) PAN/5CNT; (d) PAN/5CNT-10$CoFe_2O_4$.

the larger diameter of about 700–1000 nm may be the result of a further increase in viscosity when CNTs and $CoFe_2O_4$ nanoparticles were added into the electrospinning solution. In addition, we found that the surface of the fibers become a little rough in some places. This could be attributed to a slight aggregation of CNTs and $CoFe_2O_4$ nanoparticles within the fibers.

To investigate the crystal structures present in the nano-composite fibers, XRD patterns from PAN/5CNT and PAN/5CNT-10$CoFe_2O_4$ fibers were taken, and are shown in Fig. 7.9. For PAN/5CNT fibers, two strong diffraction peaks are observed (Fig. 7.9a). The peak, around $2\theta = 17°$, could correspond to orthorhombic PAN (110).[68] The other peak, at $2\theta = 26°$, is assigned to the (002) plane, which corresponds to the graphene layers of CNT.[69] Therefore, the appearance of these two characteristic diffraction peaks proves the existence of PAN/CNT composites. Figure 7.9b shows several strong diffraction peaks besides the two diffraction peaks shown in Fig. 7.9a. These peaks may be assigned to the (220), (311), (222), (400), (422), (511), (440) crystal planes, which are a good match with the spinel structure of $CoFe_2O_4$ (JCPDS 22-1086). The presence of the spinel structure confirms that the $CoFe_2O_4$ was synthesized by a hydrothermal process. The average particle

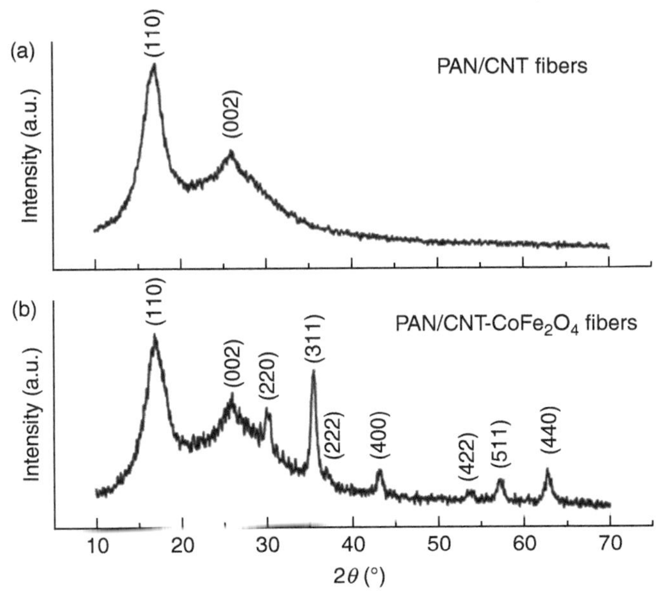

7.9 XRD patterns of (a) PAN/5CNT fibers and (b) PAN/5CNT-10$CoFe_2O_4$ fibers.

size of $CoFe_2O_4$ was further estimated using the Scherrer formula, and the calculation result based on the peak indexed (311) was about 17 nm. From the XRD pattern shown in Fig. 7.9b, we conclude that the both CNTs and nanoparticles were in agreement with the PAN matrix.

The presence of CNTs inside the nano-composite fibers was further confirmed using Raman spectroscopy. The obtained spectrum of PAN/5CNT-10$CoFe_2O_4$ fibers is shown in Fig. 7.10, where three obvious bands are observed. A comparison with the spectrum of PAN fibers (shown in the inset) confirms that the band centered at 2240 cm^{-1} comes from PAN, and corresponds to the nitrile group (–CN).[70] The other two bands that could be assigned to D and G bands come from CNTs – referring to the spectra in Fig. 7.6. The Raman spectra (as shown in Fig. 7.10) thus prove the incorporation of CNTs into the PAN matrix. Compared with Fig. 7.6, we also note that there is a slight shift of Raman bands and an increase of $R = I_D/I_G$ for the CNTs incorporated in PAN, which may be attributed to the interactions between CNTs and PAN.[71]

TEM observations were performed to study the distribution of CNTs and $CoFe_2O_4$ nanoparticles inside the nano-composite fibers. For PAN/10$CoFe_2O_4$ fibers (Fig. 7.11a), it could be clearly observed that $CoFe_2O_4$ nanoparticles were embedded in PAN, although the distribution was not homogeneous due to the aggregation of $CoFe_2O_4$ nanoparticles. The nano-scale size, high concentration, and lack of subsequent surface modification of these

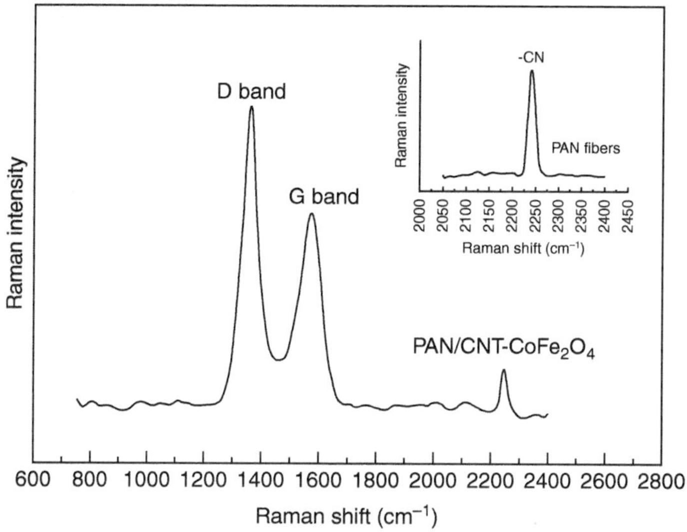

7.10 Raman spectrum of PAN/5CNT-10$CoFe_2O_4$ fibers; Inset shows the Raman spectrum of PAN fibers.

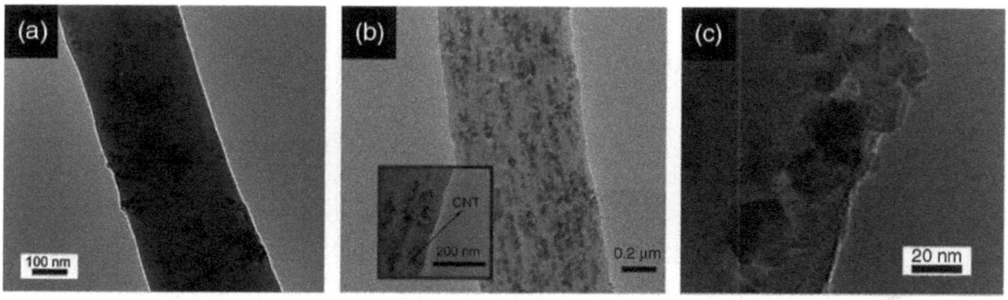

7.11 TEM images of (a) PAN/10$CoFe_2O_4$ fiber; (b) PAN/5CNT-10$CoFe_2O_4$ fiber and (c) $CoFe_2O_4$ nanoparticles on the surface of composite fiber.

as-prepared $CoFe_2O_4$ particles may be the reasons for aggregation during the electrospinning process.

Figure 7.11b shows a PAN/5CNT-10$CoFe_2O_4$ fiber. The $CoFe_2O_4$ nanoparticles inside the PAN fiber exhibit a general ribbon arrangement along the fiber axis, which is very different from the distribution of $CoFe_2O_4$ nanoparticles observed in Fig. 7.11a. In contrast to the obvious presence of $CoFe_2O_4$ nanoparticles, the CNTs are hardly seen. This may be partly attributed to the relatively larger size of the fiber. Here, we further assume that the $CoFe_2O_4$ nanoparticles absorb onto the CNTs, and many of the CNTs were encapsulated by these $CoFe_2O_4$ nanoparticles. Those functional groups (such as carboxyl) on the surface of CNTs that were introduced by nitric acid treatment were considered to serve as attachment points to anchor the other nanoparticles.[72] CNTs have been observed to be well arranged along the axis in polymer fibers by electrospinning.[73,74] This also explains the ribbon arrangement of $CoFe_2O_4$ nanoparticles inside the PAN fibers. As shown in the inset in Fig. 7.11b, a CNT partially unencapsulated by $CoFe_2O_4$ nanoparticles is observed, further supporting this assumption. Due to the occasional defects appearing on the surface of the nano-composite fibers, incomplete embedding of $CoFe_2O_4$ nanoparticles inside the PAN fibers is observed – as shown in Fig. 7.11c. The high magnification image clearly shows the synthesized $CoFe_2O_4$ nanoparticles, with a particle size about 10–25 nm, which is in agreement with the estimation result from XRD pattern.

The incorporation of CNTs into a polymer matrix can increase conductivity,[75,76] suggesting that these composites may have electromagnetic shielding applications. To test whether the above-fabricated PAN/CNT-$CoFe_2O_4$ fibers would exhibit shielding properties, EMISE of PAN/10CNT-5$CoFe_2O_4$ fiber sheets was evaluated. Shielding effectiveness (SE) is defined as the attenuation of electromagnetic waves, and is expressed as:

$$SE = 10\log\frac{P_O}{P_T}$$

where P_O and P_T respectively refer to the power of incident and transmitted electromagnetic waves through a shielding material.[76,77]

Figure 7.12a shows the shielding curve of a PAN/10CNT-5$CoFe_2O_4$ fiber sheet over 8–12 GHz, and a curve from a PAN /5CNT-10 $CoFe_2O_4$ fiber sheet is presented for comparison. The areal density was controlled at about 9.7 mg/cm^2 for both samples. Compared with the PAN/5CNT-10$CoFe_2O_4$ fiber sheet, the PAN/10CNT-5$CoFe_2O_4$ fiber sheet exhibits a higher SE. This indicates that the SE of the nano-composite fibers depends on the concentration of CNTs, and increases with increased fiber CNT content. Conductivity is considered an important factor for shielding, and a higher conductivity should bring a higher SE. Therefore, the increased SE observed in the nano-composite fibers could be related to improvements in conductivity owing to the formation of a CNT conductive network. No obvious SE effect from the $CoFe_2O_4$ nanoparticles is observed over this frequency range. In Fig. 7.12b, the SE curves (PAN/10CNT-5$CoFe_2O_4$ fiber sheets) corresponding to different areal densities (9.7, 19.4, 29.1, 38.8 mg/cm^2) are arranged to confirm the effect of areal density.

It is observed that the SE significantly increases with increased areal density. For the sample with an areal density of 38.8 mg/cm^2, an SE of about 3.9 dB is achieved at 12 GHz. From the results of our SE evaluation, we can conclude that the SE of PAN/CNT-$CoFe_2O_4$ fibers depends on CNT concentration and areal density: higher CNT concentration and areal density

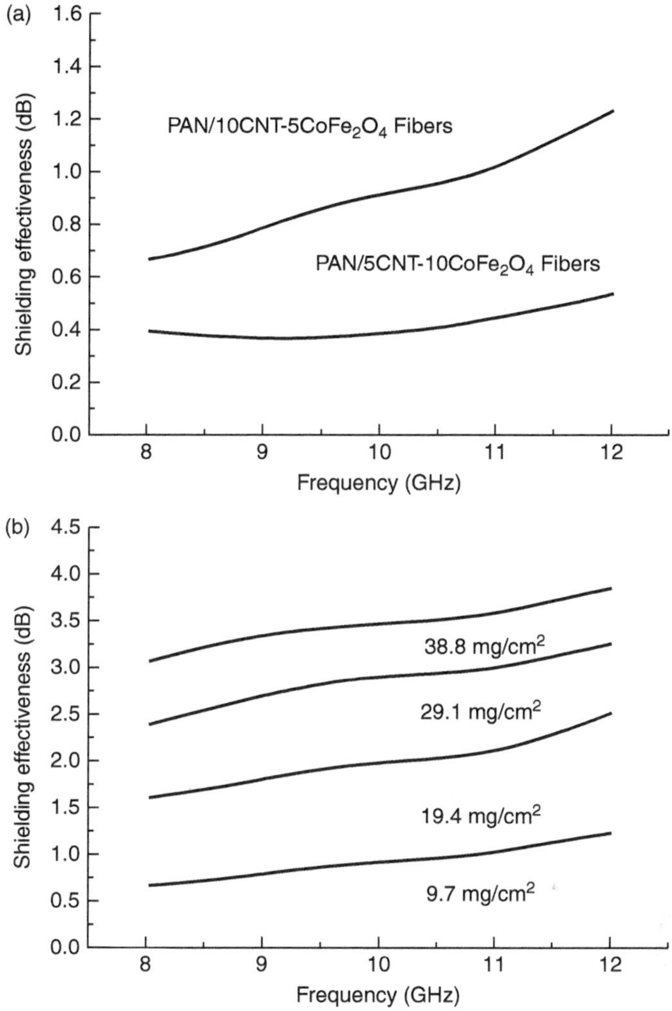

7.12 SE curves of (a) PAN/5CNT-10$CoFe_2O_4$ and PAN/10CNT-5$CoFe_2O_4$ fiber sheets; (b) PAN/10CNT-5$CoFe_2O_4$ fiber sheets with different areal density.

yield higher SE. Thus, nano-composite fibers have possible applications in electromagnetic shielding.

To illustrate the fibers' magnetic properties, the room temperature magnetization curves of PAN/5CNT-10$CoFe_2O_4$ fibers and as-prepared $CoFe_2O_4$ nanoparticles are shown in Fig. 7.13, and typical hysteresis loops are observed for both. For the PAN/5CNT-10$CoFe_2O_4$ fibers, the saturation magnetization (Ms) and coercivity (H_C) are 4.7 emu/g and 0.8 kOe, respectively. The magnetic properties of the PAN/5CNT-10$CoFe_2O_4$ fibers arise from their

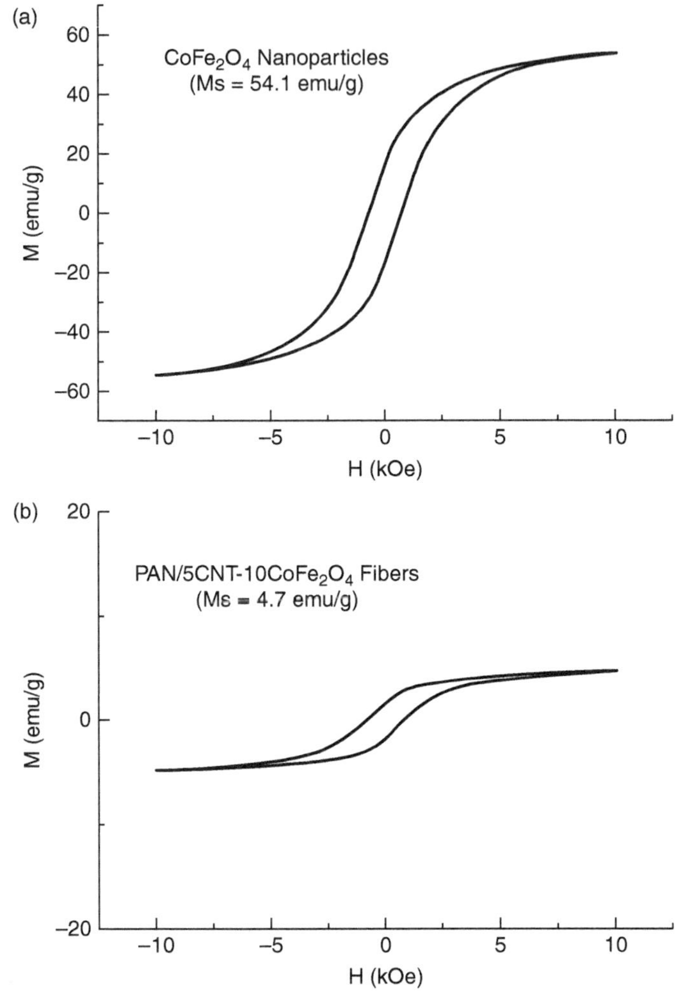

7.13 Magnetization hysteresis loops of (a) $CoFe_2O_4$ nanoparticles, and (b) PAN/5CNT-10$CoFe_2O_4$ fibers.

incorporated $CoFe_2O_4$ nanoparticles. Based on the $CoFe_2O_4$ loading in PAN/5CNT-10$CoFe_2O_4$ fibers, the calculated value of pure $CoFe_2O_4$ nanoparticles is about Ms = 54 emu/g, which is very close to the experimental value of Ms = 54.1 emu/g indicated by the curve taken from $CoFe_2O_4$ nanoparticles. This indicates that the Ms of the nano-composite fibers is well matched with the proportion of $CoFe_2O_4$ nanoparticles in the nano-composite fibers, showing that the magnetic properties of these nano-composite fibers can be accurately tailored by adjusting their $CoFe_2O_4$ nanoparticle loading.

7.6 Conclusion

In the first step of this work, $CoFe_2O_4$ nanoparticles with a particle size of about 10–25 nm were synthesized using a simple hydrothermal process. $CoFe_2O_4$/PAN nanofibers with various $CoFe_2O_4$ loadings (1 wt.%, 4 wt.%, 7 wt.%, 10 wt.%) were prepared by electrospinning, and the fiber diameter was controlled to a distribution of between 200 nm and 400 nm. The presence of $CoFe_2O_4$ nanoparticles in PAN nanofibers was confirmed by XRD characterization.

PAN nano-composite fibers containing CNTs and $CoFe_2O_4$ nanoparticles were fabricated by electrospinning. The presence of $CoFe_2O_4$ nanoparticles and CNTs in the PAN matrix was confirmed by XRD patterns and Raman spectra. SEM showed the regular morphologies of the fabricated nano-composite fibers, and TEM observation revealed the distribution of CNTs and $CoFe_2O_4$ nanoparticles inside the nano-composite fibers. As a multifunctional material, the nano-composite fibers achieved an EMI SE of about 3.9 dB, and exhibited magnetic properties with an Ms of 4.7 emu/g for the developed PAN/5CNT-10$CoFe_2O_4$ fibers. This property can be accurately tailored by adjusting $CoFe_2O_4$ nanoparticle loading.

These results show that electrospinning technology can be used effectively in producing various fiber materials at a nano-scale. Multifunctional nano-composite fibers can also be fabricated; these fibers potentially have extensive industrial applications.

7.7 References

1. S. H. Tan, R. Inai, M. Kotaki and S. Ramakrishna (2005), *Polymer*, **46**, 6128.
2. J. H. Park, B. S. Kim, Y. C. Yoo, M. S. Khil and H. Y. Kim (2007), *J Appl Polym Sci*, **107**, 2211.
3. J. Doshi and D. H. Reneker (1995), *J Electrostat*, **35**, 151.
4. K. Wei, T. Ohta, B. S. Kim, K. H. Lee, M. S. Khil, H. Y. Kim and I. S. Kim (2010). *Polym Adv Technol*, **21**, 746.
5. F. Liu (2012), Development of functional polyvinyl alcohol nanofibers by electrospinning technology, Doctoral thesis in Shinshu University.
6. H. Fong, I. Chun and D. H. Reneker (1999), *Polymer*, **40**, 4585.
7. Y. M. Shin, M. M. Hohman, M. P. Brenner and G. C. Rutledge (2001), *Polymer*, **42**, 9955.
8. H. Fong, W. Liu, C. S. Wang and R. A. Vaia (2000), *Polymer*, **43**, 775.
9. P. P. Tsai, H. Schreuder-Gibson and P. Gibson (2002), *J Electrostat*, **54**, 333.
10. P. Gibson, H. Schreuder-Gibson and D. Rivin (2001), *Colloid Surface A*, **187–188**, 469.
11. E. R. Kenawy, G. L. Bowlin, K. Mansfield, J. Layman, D. G. Simpson, E. H. Sanders and G. E. Wnek (2002), *J Control Release*, **81**, 57.
12. Y. K. Luu, K. Kim, B. S. Hsiao, B. Chu and M. Hadjiargyrou (2003), *J Control Release*, **89**, 341.

13. K. Suzuki (2011), Fabrication and characterization of oriented nano fiber nonwoven by electrospinning method, Master thesis at Shinshu university.
14. R. Dersch, A. Greiner, M. Steinhart and J. H. Wendorff (2003), *DWI Report*, **8**, 127.
15. A. Greiner (2005), *Lecture on 'Electrospinning for Biomedical Applications'*, DWI Aachen, Germany.
16. H. Thomas, E. Heine, R. Wollseifen, C. Cimpeanu and M. Möller (2005), *Int Nonwovens J*, **14**(3), 12.
17. T.H. Grafe and K.M. Graham (2003), Nanofiber webs from electrospinning, *Nonwovens in Filtration, 5th International Conference, Stuttgart*, Germany, 11–13 March.
18. H. Schreuder-Gibson, P. Gibson, K. Senecal, M. Senett, J. Walker, W. Yeomans, D. Ziegler and P.P. Tsai (2002), *Adv Mater*, **34**, 44.
19. L. Huang, R.A. McMillan, R.P. Apkarian, B. Pourdeyhimi, V. Conticello and E.L. Chaikof (2000), *Macromolecules*, **33**, 2989.
20. G. Verreck, I. Chun, J. Rosenblatt, J. Peeters, A. Van Dijck, J. Mensch, M. Noppe and M.E. Brewster (2003), *J Control Release*, **92**, 349.
21. A. Frenot and I.S. Chronakis (2003), *Colloid Interf Sci*, **8**, 64.
22. P. Gibson, H. Schreuder-Gibson and D. Rivin (2001), *Colloids Surf A: Physicochem Eng Asp*, **187–188**, 469.
23. M. Jacobsen (1991), *Nonwovens Industry*, **36**. cf. also: Chemiefasern/Textilindustrie **39/91**, 868 (1989).
24. E.R. Kenawy, G.L. Bowlin, K. Mansfield, J. Layman, D.G. Simpson, E.H. Sanders and G.E. Wnek (2002), *J Control Release*, **81**, 57.
25. L. Larrondo and R. St. John Manley (1981), *J Polym Sci: Polym Phys Ed*, **19**, 909.
26. L. Larrondo and R. St. John Manley (1981), *J Polym Sci: Polym Phys Ed*, **19**, 921.
27. L. Larrondo and R. St. John Manley (1981), *J Polym Sci: Polym Phys Ed*, **19**, 933.
28. G.I. Taylor (1964), *Proc R Soc Lond Ser A*, **280**, 383.
29. G.I. Taylor (1969), *Proc R Soc Lond Ser A*, **313**, 453.
30. M. Shin, M.M. Hohman, M.P. Brenner and G.C. Rutledge (2001), *Appl Phys Lett*, **78**, 1149.
31. M.M. Hohman, M. Shin, G. Rutledge and M.P. Brennera (2001), *Phys. Fluids*, **13**(8), 2201.
32. P. Dalton, T. Kuenzel, D. Klee and M. Moeller (2004), *J Mey Eur Cell Mater*, **7**(1), 52.
33. A Weghmann (1981), *Schriftenreihe des Deutschen Wollforschungsinstitutes*, **85**, 314.
34. K. Schmidt (1980), *Melliand Textilber*, **61**(6), 495.
35. D. Luzhansky (2003), *INDA/TAPPI*, Baltimore, MD, 16–18 September.
36. Air Filtration, Liquid Filtration (2013), http://www.hovo.com.
37. Nanospider (2013), http://www.elmarco.cz and http://www.nanospider.c z/aplikace.php?kategorie=1&h.
38. X. D. Jin, Q. Q. Ni, Y. Q. Fu, L. Zhang, T. Natsuki (2012), *Polymer Composites*, **33**, 317–323.

39. J. Dexmer, C. M. Leroy, L. Binet, V. Heresanu, P. Launois, N. Steunou, C. Coulon, J. Maquet, N. Brun, J. Livageand and R. Backov (2008), *Chem Mater*, **20**, 5541.
40. H. Guo and S. Tao (2007), *Sens Actuators B*, **123**, 578.
41. S. Y. Yeo and S. H. Jeong (2003), *Polym Int*, **52**, 1053.
42. S. Y. Yeo, H. J. Leeand and S. H. Jeong (2003), *J Mater Sci*, **38**, 2143.
43. Y. Li and T. Chung (2008), *Microporous Mesoporous Mater*, **113**, 315.
44. A. Alshebani, M. Pera-Titus, E. Landrivon, Th. Schiestel, S. Miachon and J.-A. Dalmon (2008), *Microporous Mesoporous Mater*, **115**, 197.
45. L. Y. Jiang and T. S. Chung (2006), *AIChE J*, **52**, 2898.
46. H. Wolf, Z. Pajkic, T. Gerdes and M. Willert-Porada (2009), *J Power Sources*, **190**, 157.
47. H. Lu, D. Li, K. Sun, Y. Li and Z. Fu (2009), *Solid State Sci*, **11**, 982.
48. M.J. Uddin, F. Cesano, F. Bonino, S. Bordiga, G. Spoto, D. Scarano and A. Zecchina (2007), *J Photochem Photobiol A*, **189**, 286.
49. M. Jin, X. Zhang, A. V. Emeline, Z. Liu, D. A. Tryk, T. Murakami and A. Fujishima (2006), *Chem Commun*, (43), 4483–85.
50. D. Zhang, A. B. Karki, D. Rutman, D. P. Young, A. Wang, D. Cocke, T. H. Ho and Z. Guo (2009), *Polymer*, **50**, 4189.
51. D. Chen, T. Liu, X. Zhou, W. C. Tjiu and H. Hou (2009), *J Phys Chem B*, **113**, 9741.
52. H. Zhang, H. Song, H. Yu, X. Bai, S. Li, G. Pan, Q. Dai, T. Wang, W. Li, S. Lu, X. Ren and H. Zhao (2007), *J Phys Chem C*, **111**, 6524.
53. S. Wang, C. Wang, B. Zhang, Z. Sun, Z. Li, X. Jiang and X. Bai (2010), *Mater Lett*, **64**, 9.
54. B. Wang, Y. Sun and H. Wang (2010), *J Appl Polym Sci*, **115**, 1781.
55. T. Song, Y. Zhang, T. Zhou, C. T. Lim, S. Ramakrishna and B. Liu (2005), *Chem Phys Lett*, **415**, 317.
56. H. G. Chae, T. V. Sreekumar, T. Uchida and S. Kumar (2005), *Polymer*, **46**, 10925.
57. L. Vivien, P. Lançon, D. Riehl, F. Hache and E. Anglaret (2002), *Carbon*, **40**, 1789.
58. O. Meincke, D. Kaempfer, H. Weickmann, C. Friedrich, M. Vathauer and H. Warth (2004), *Polymer*, **45**, 739.
59. M. H. Khedr, A. A. Omar and S. A. Abdel-Moaty (2006), *Colloids Surf A*, **281**, 8.
60. Y. Ju, J. Park, H. Jung, S. Cho and W. Lee (2008), *Mater Sci Eng B*, **147**, 7.
61. E. J. Ra, K. H. An, K. K. Kim, S. Y. Jeong and Y. H. Lee (2005), *Chem Phys Lett*, **413**, 188.
62. I. Chen, C. Wang and C. Chen (2010), *J Phys Chem C*, **114**, 13532.
63. N. G. Sahoo, S. Rana, J. W. Cho, L. Li and S. H. Chan (2010), *Prog Polym Sci*, **35**, 837.
64. X. Xie, Y. Mai and X. Zhou (2005), *Mater Sci Eng R*, **49**, 89.
65. T. McNally, P. Pötschke, P. Halley, M. Murphy, D. Martin, S. E.J. Bell, G. P. Brennan, D. Bein, P. Lemoine and J. P. Quinn (2005), *Polymer*, **46**, 8222.
66. S. Osswald, M. Havel and Y. Gogotsi (2007), *J Raman Spectrosc*, **38**, 728.
67. P. Heikkilä and A. Harlin (2009), *Express Polym Lett*, **3**, 437.
68. J. Chen, Z. Li, D. Chao, W. Zhang and C. Wang (2008), *Mater Lett*, **62**, 692.

69. G. Wang, X. Shen, J. Yao, D. Wexler and J. Ahn (2009), *Electrochem Commun*, **11**, 546.
70. H. Hou, J. J. Ge, J. Zeng, Q. Li, D. H. Reneker, A. Greiner and S. Z. D. Cheng (2005), *Chem Mater*, **17**, 967.
71. H. Zhang, L. Xu, F. Yang and L. Geng (2010), *Carbon*, **48**, 688.
72. K. A. Wepasnick, B. A. Smith, K. E. Schrote, H. K. Wilson, S. R. Diegelmann and D. H. Fairbrother (2011), *Carbon*, **49**, 24.
73. J. J. Ge, H. Hou, Q. Li, M. J. Graham, A. Greiner, D. H. Reneker, F. W. Harris and S. Z. D. Cheng (2004), *J Am Chem Soc*, **126**, 157–54.
74. K. Saeed and S. Park (2010), *J Polym Res*, **17**, 535.
75. T. Yang, D. Wu, L. Lu, W. Zhou and M. Zhang (2011), *Polym Compos*, **32**, 1280.
76. Y. Li, C. Chen, S. Zhang, Y. Ni and J. Huang (2008), *Appl Surf Sci*, **254**,57–66.
77. J. S. Im, J. G. Kim, S. Lee and Y. Lee (2010), *Colloids Surf. A*, **364**, 151.

8
Integrated composite spinning (ICS)

W. OXENHAM, North Carolina State University, USA

DOI: 10.1533/9780857099174.2.149

Abstract: The chapter reviews ICS, which is the creation of a yarn by bonding staple fibers to a filament core using molten polymer as the bonding agent. An overview of the development of the concept through to a commercial machine is presented, including the optimization of the key components. The results of independent studies on the system are summarized, including analyses of yarn structure and properties. The conclusion offers thoughts on why a system with so much promise was not adopted by the industry.

Key words: integrated composite spinning (ICS), Bobtex, hybrid yarns.

8.1 Introduction

Integrated Composite Spinning (ICS) is based on the concept of combining filament yarns with staple fibers to produce a so-called 'composite yarn' (Denton and Daniels, 2002). There have been many approaches to spinning composite yarns, including the creation of core spun yarns on a ring frame (Jeddi *et al.*, 2009) and wrap-spun yarn using hollow-spindle technology (Xie *et al.*, 1986; Jiang and Chen, 2012). Techniques utilizing electrostatic charging of the filaments, prior to their combination with staple fibers, were proposed to better integrate the filament and staple components (Jou *et al.*, 1996). Recently, core spun yarns, consisting of a filament core covered with staple fibers, have also been created using systems such as rotor spinning (Nield and Ali, 1977), friction spinning, and jet spinning (Ortlek and Ulku, 2007). All of these techniques rely on twist (or false twist) to provide structural integrity to the resultant yarn. An alternative approach was to use an adhesive to stick the staple fibers to each other. In commercial applications the adhesive used was water soluble and this was removed after fabric formation, yielding a fabric with good aesthetics and performance (Selling, 1971).

What differentiated ICS from other composite yarn formation systems was that the process was designed to integrate a staple fiber component into 'semi-molten' polymer. The 'semi-molten' polymer could be an extruded filament that had not fully cooled, and staple fibers were added to the still 'tacky' surface. The favored technology was to start with a filament yarn and coat it with molten polymer, and then to add staple fibers to the surface before the polymer had cooled. The technology was invented and brought to commercial realization by the Bobtex Corporation (of Montreal, Canada) and marketed as the Bobtex ICS system (Mohamed, 1973). The system was initially launched during the early 1970s, which was the era when many new technologies were being introduced into staple spinning, most of which had the goal of replacing ring spinning as the standard approach to staple yarn production.

The concept aroused a lot of excitement, because of its high production speed compared to all other staple spinning competitors, and several installations underwent extensive industrial trials. Additionally, a number of research papers were published on different aspects of the technology, ranging from the structure and properties of yarns and fabrics through to discussions of the potential economic benefits of the system. Ultimately, the system was not a major commercial success and, although the technology was viewed as revolutionary when it was first introduced, it is now regarded as a historical curiosity.

This chapter will review the development of the process, the influence of processing parameters on the structure and properties of yarns and fabrics, the strengths and weaknesses of the system, and a comment on whether there is a potential for such a system in an industry which now has a greater focus on technical, performance, and medical textiles.

8.2 The background and history of integrated composite spinning (ICS)

The following sections briefly describe the history of the Bobtex Integrated Composite Spinning Process and the transition from a concept into a machine which was capable of creating yarns with staple fibers at significantly higher speeds than any machine available at that time or indeed today

8.2.1 Concept originators

The development of the ICS technology can be traced back to a suite of patents filed by the father and son team of Emilian and Andrew J. Bobkowicz, who founded the Bobtex Corporation in Montreal, Canada. This team (Figs 8.1 and 8.2) had many patents over a wide range of textile related topics, and history may remember them as great inventors but not so great as entrepreneurs.

8.1 Emilion Bobkowicz at the offices of Bobtex Corporation. Photograph provided by Bobtex Corporation to the author, 15 September 1975.

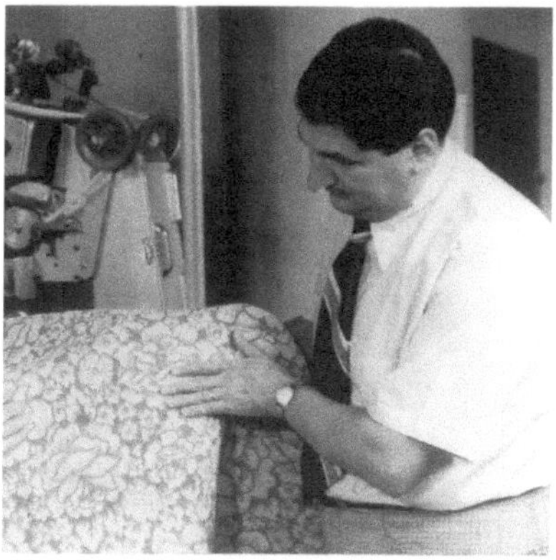

8.2 A. J. Bobkowicz examining upholstery fabric made with Bobtex ICS filling. Photograph provided by Bobtex Corporation to the author, 23 September 1975.

8.2.2 Prototype development

The technology used to manufacture ICS yarns evolved over several years and different prototype machines. While the concept of the final version may have been part of the original vision, it is evident that several alternative approaches were also tried. The key feature of the earlier proposals was to embed staple fibers into an extruded, but not fully cooled, thermoplastic

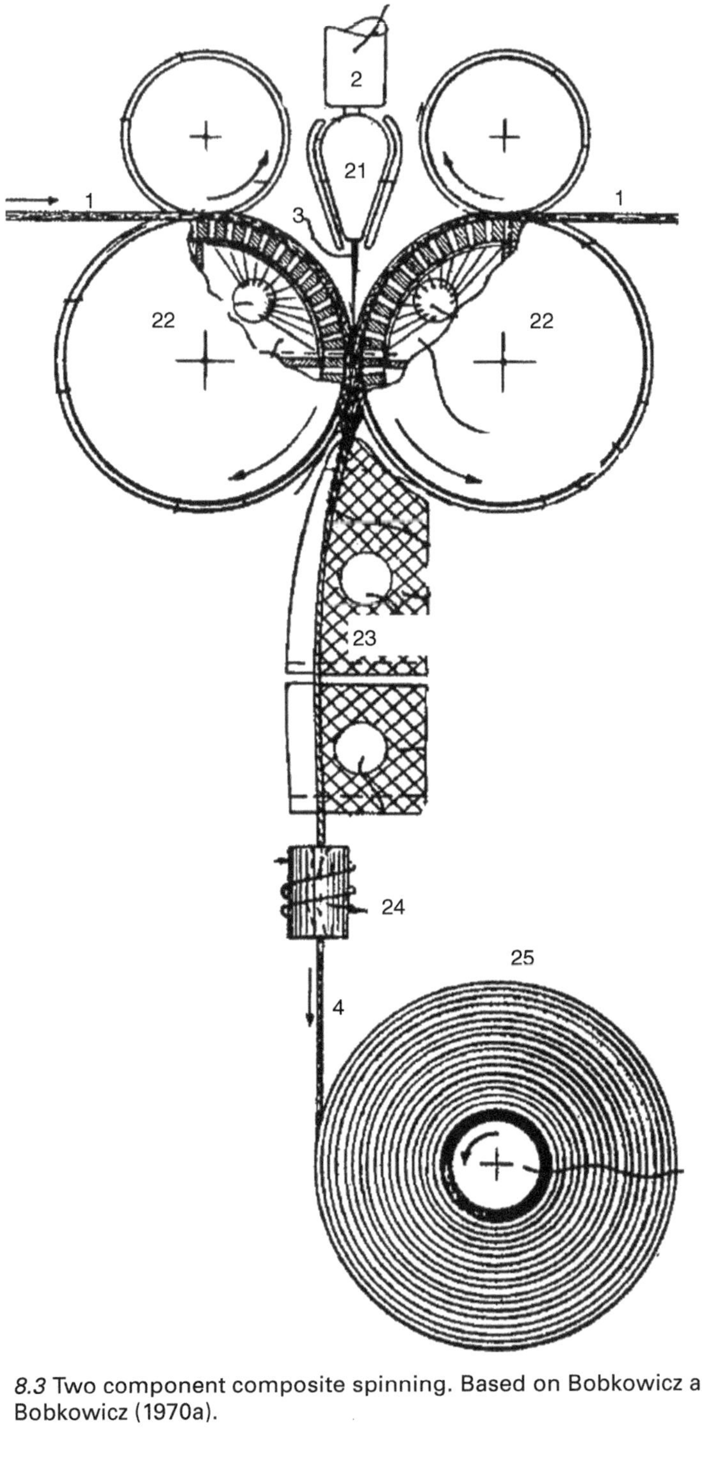

8.3 Two component composite spinning. Based on Bobkowicz and Bobkowicz (1970a).

polymer. Since the polymer had not cooled, the staple fiber adhered to the 'sticky' surface of the polymer. The fibers were then further embedded into the polymer by the movement of the polymer surface, which was induced using a false twist device in the threadline (Bobkowicz and Bobkowicz, 1970a). The product of this machine was known as a two component yarn, consisting of staple fibers embedded in an extruded thermoplastic polymer. The potential to include a filament yarn as a core and thus produce a three component yarn was indicated, but this is best regarded as a further development, or refinement of the technology. What differentiated the 'prototype' from later developments was the way the polymer and fibers were assembled.

A schematic of the prototype is shown in Fig 8.3. Polymer chips in a hopper (2) were fed to a melt extrusion unit (21) resulting in the production of 'filament' (3). The filament passed between a pair of nip rollers (called laminating rollers), which were covered with staple fibers. The nip rollers (22) not only provided a method of squeezing the staple fibers into the surface of the filament but could also act as draw rollers, imparting orientation (and improved tensile properties) to the 'filament.' The staple fibers were typically fed to the machine as sliver (1) and, in earlier versions, the sliver was drafted by a roller drafting system into a thin web. The laminating rollers were also perforated, and suction through the rollers held the fibers on the roller surface before they were transferred to the filament. The composite structure now passed through a false twist device (24), which caused the fibers to become further embedded into the polymer. Because the polymer component was still not fully cooled, the false twisting created surface shear and the resultant yarn, which was cooled (23) as it passed to the winder, exhibited the surface character of a twisted yarn. The resultant composite yarn (4) was taken up onto a high speed winding unit (25).

A proposed alternative to the above approach was to create a polymer film/fiber tape that could then be slit, hot-drawn, and twisted to create the yarn, but this was not further developed (Bobkowicz and Bobkowicz, 1970b).

8.3 The development of ICS technology

While the concept of creating composite structures had earlier been proposed by Bobkowicz – indeed the idea had been indicated in patents from 1965 – these had been associated with the terminology 'textured filament yarn' (Bobkowicz and Bobkowicz, 1967), and the first patent specifically relating to 'Composite Spinning' was granted in 1974 (Bobkowicz and Bobkowicz, 1974a).

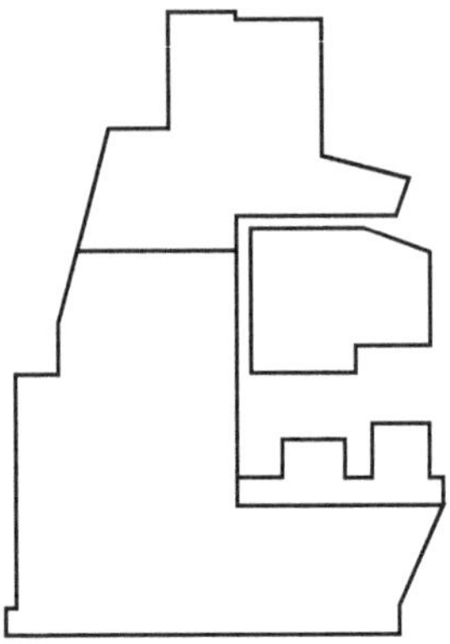

8.4 Construction 'modules' of Bobtex ICS machine. Based on a diagram in Mohamed (1973).

8.3.1 The machine as an assembly of specific components

The machine was regarded as an assembly of several modules, which are shown in Fig. 8.4, and each module was specially developed for the creation of composite yarns (Mohamed, 1973). The extrusion unit was designed to be capable of handling many polymer types, and claimed to be 'electrically heated with automatic temperature control and simple setting of the temperature for the resin. Built to be operated by textile workers with no experience in extrusion of polymers, ease of cleaning and maintenance has been incorporated in the design of the extruder' (ibid.). The spinning unit was where the staple fibers were joined to the polymer (or polymer-coated core yarn), and the fiber feed system that was ultimately adopted was an opening roller system (sometimes called combing roller). The twisting unit was a bush friction twister capable of speeds up to 25 000 revolutions per minute. The winding module had two winding heads for each position to enable continuous running, and was capable of producing 3.5 kg packages at 600 m/min. The drive module housed the motors and controllers for the machine.

8.3.2 Application of technology in achieving functional components

The schematic of the Bobtex ICS machine is shown in Fig. 8.5 and a picture of the Bobtex Mk1 ICS is given in Fig. 8.6. From the diagram, Fig. 8.5, it can be seen that a yarn, (called the core, carrier or filament) was taken from a supply package and passed through an extrusion unit where it was coated with molten polymer (resin). The coated yarn passed down to a laminating

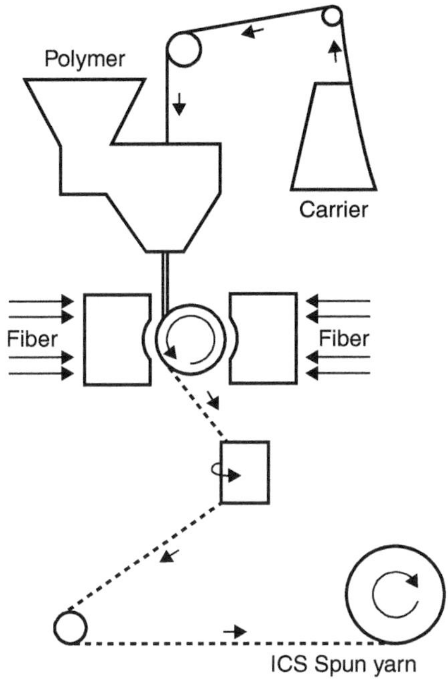

8.5 Schematic diagram of Bobtex Mk1 ICS machine. Redrawn from illustration in Riordon and Bobkowicz (1974).

roller covered with staple fibers, and these adhered to the still tacky polymer. On its way to a winding unit the yarn was subjected to a false twisting action, and this consolidated the structure and drove the fibers into the polymer. Because of the plastic nature of the polymer, the net effect was a yarn that had the characteristics of a twisted structure.

The machine, Fig. 8.6, consisted of two positions, mounted back to back, each producing a yarn. A single extrusion unit fed molten polymer to nozzles on both sides of the machine. Figure 8.5 shows that two slivers and a core yarn were fed into each side of the machine. The Mk 1 machine did have the possibility of producing two component yarns, but was primarily designed to create three component ICS using a filament core (carrier). To this end, there was significant development in the extrusion unit, particularly in the design of the die and nozzle arrangements. For a three component yarn, the filament core passed through the nozzle where the polymer was extruded as an annulus to give complete coverage to the surface of the core.

If a two component yarn was required, the jet could be simpler; however, there were some limitations. It had been found that it was better to use a core when lower quality, shorter fibers were used, but for other applications the core could be eliminated. When not using a core there was the

8.6 Bobtex Mk1 ICS Machine. Photograph provided by Bobtex Corporation to author, 15 September 1975.

possibility of drawing the extruded polymer in order to a give a stronger product; however, this could present problems. It was found that at speeds beyond 150 m/min there were instabilities in the threadline between the nozzle and the laminating rollers, which, as indicated above, were also acting as draw rolls. A solution to this problem was to use a 'spray extrusion technique' in which the use of compressed air and a modified nozzle blew a fine continuous polymer stream to the nip of the laminating rollers. This not only eliminated the problem of instability, and hence enabled much higher speeds, but it also better integrated the polymer with the staple fiber and produced a better yarn.

Figure 8.7 (Bobkowicz and Bobkowicz, 1974a) shows a much simplified schematic of the nozzle design. If a two component yarn were being produced the molten polymer would pass down the central orifice and air would be blown down the 'annulus,' indicated in the diagram as B. The air flow improved the spinning stability and resulted in a better yarn quality. For three component yarn, the yarn core would pass through the central orifice, and molten polymer would issue through B to provide the coating, to which the staple component would later adhere.

In the case of a three component yarn there were also some problems associated with the appearance and staple fiber cover of the final yarn and these were improved by the use of an additional vortex at the exit of the

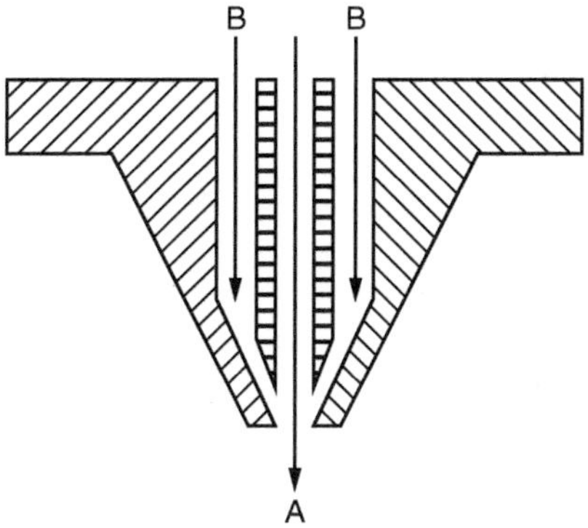

8.7 Schematic of nozzle designs based on proposals made by Bobkowicz and Bobkowicz (1974a).

nozzle. Since for a three component yarn there was no drawing between the nozzle and the laminating roller, the covered polymer was caused to swirl and this created better cover and appearance in the final yarn (Bobkowicz and Bobkowicz, 1974b).

The staple fiber opening system consisted of an opening roller, similar to that found in a rotor spinning machine. It is notable, however, that the productivity of a Bobtex unit is significantly greater than a rotor spinning machine and thus the opening roller is working under much more demanding conditions (heavier slivers being processed at much higher throughput speeds). Optimization of the opening roller resulted in the use of pins rather than card wire for the roller covering, since this was claimed to give better opening with less fiber damage (Lord, 1973).

The streams of fibers that issued from the openers were collected on a perforated roller through which air was sucked to hold the fibers on the roller. This was the so-called laminating roller, since this was also contacted by the tacky polymer (or the polymer coating of the core yarn), and the fibers were squeezed into the polymer. There was such a unit at either side of the threadline, and the laminating rollers formed a nip for drawing when making a two component yarn. Later machines, which were designed for three component yarns, used only one laminating roller, since the three component yarn could not be drawn. There were some issues with the laminating roller, since at higher speed the centrifugal force could overcome the suction force and fiber could fly off. Modifications to the laminating roller, which included additional suction slots, overcame this problem and permitted higher speeds (Bobkowicz and Bobkowicz, 1975a). Table 8.1 shows examples of yarns produced and the capabilities of the improved grooved laminating roller.

Table 8.1 Examples of yarn production parameters and resultant properties

Fibers (F)	Carrier (C)	Resin (R)	Count (tex)	Tenacity (gf/tex)	Composition (F/C/R%)	Fiber feed (ft/min)	Grooved roller speed (ft/min)	Twist applied (turns/in.)
Polypropylene 2.5 in. 15 denier Sliver 4134 tex	Polypropylene Ribbon 40 tex	Polyethylene M.I. 150 M.P. 300°F	179	10.1	34/22/44	4.9	330	22
Cotton 15/16 in., Sliver 4400 tex	Glass multifilament 33 tex	Polyethylene M.I. 150 M.P. 300°F	114	15.6	27/28/45	2.1	300	54
Polyester 1.5 in. 1.5 denier Sliver 4000 tex	Polyester multifilament 15 tex	Polyester i.v. = 70 M.P. 345°F	72	12.1	37/2[illegible]/42	9.3	1400	74
Polyester/cotton waste (5/8 in.) Web 4400 tex	Polyester multifilament 15 tex	Polyethylene M.I. 150 M.P. 300°F	46	11.5	45/25/30	7.3	2000	40
Rayon 1 9/16 in. 1.5 denier Sliver 3015 tex	Nylon multifilament 11 tex	Polyethylene M.I. 150 M.P. 300°F	46	10.8	34/23/43	21.3	4100	43

The element that consolidated the yarn was the false twisting unit, which was a bush friction twisting device, similar to those used on texturing machines at that time. This could be driven up to 25 000 revolutions per minute but, because of the large ratio of the bush diameter to the yarn diameter, the yarn rotated at much higher speeds. The amount of false twist inserted greatly affected the structure of the yarn and in particular the consolidation, which in turn determined the amount of staple fiber remaining on the surface. Since the polymer was still 'soft' there was slippage of the sheath around the core in three component yarns, and plastic deformation of the polymer in a two component yarn. This resulted in the false twist creating a yarn that appeared to have real twist, and the magnitude of this effect was obviously influenced by the level of false twist.

Two further improvements to the basic concept were proposed. The first was a broad-based patent based on the development of an open-end spinning technology called aerodynamic break spinning (ABS) but which utilized differential twisting to create a novel yarn structure from staple fibers at very high speeds (claims of potentially 3000 m/min). Within the general claims, there is also the potential to include a filament core (to make a core spun yarn) or to incorporate a filament covered with molten polymer (Bobkowicz, 1976). When viewed against the ICS system, the proposed approach was much more complex and was not commercialized.

Table 8.2 ICS yarn examples using FOY and POY filament carrier yarn

Count (tex)	75	80	95	89	190	140
Fiber (%)	W 55	W 57	A 45	A 44	V 60	W/MA 27
Polymer (%)	PPA 23	PPA 22	PPA 19	PPA 19	PPA 14	SA 49
Filament (%)	PETP 22	PETP 21	PETP 36	PETP 37	PETP 26	Glass 24
	1 × POY	1 × FOY	2 × POY	2 × POY	3 × POY	1
Speed (ft/min)	1450	2034	2150	2650	2040	2040
Run tension (%)	36	65	30	40	150	50
Efficiency (%)	85	85	90	90	90	70
Adhesion Index[1]	3	1	4	5	5	5
Breaking strength (gf)	633	650	1275	870	1050	2500
Elongation (%)	30	29	21	26	46	3.3
Shrinkage[2] (%)	7	5	5	7	13	0

[1] Measure of ability to strip the fibers/polymer sheath from the carrier core filament on a scale of 1 to 5; 1 signifies easy sliding off with almost no resistance, 5 signifies no stripping, the force to strip being in excess of the total breaking strength.
[2] Shrinkage in relaxed form, dry heat 15 min at 230°F.

Abbreviations:
A- Acrylic; V-Viscose; SA-Saran Alloy; PPA-Polypropylene Alloy; W-Wool; MA-Modacrylic; PETP-Polyester PP-Polypropylene.

8.8 Bobtex Mark 3 ICS Machine. From publicity material 'Spin at 1000 m/min' supplied to the author by Bobtex International, Leicester, UK.

The second improvement was to overcome a potential problem at higher speeds of breaks, or discontinuity, in the polymer sheath surrounding the filament core. This obviously, in turn, yields a yarn with variable staple fiber cover, since fibers cannot adhere uniformly without a continuous sheath. It was demonstrated that this problem could be reduced or eliminated by using partially oriented yarn (POY), or undrawn yarn, as the core feedstock and drawing the core before or during coating. Table 8.2 shows examples of yarns produced from fully oriented yarn (FOY) and POY, and it can be observed that the use of POY (and multiple cores) gives significant improvements in adhesion between the core and the polymer, and general improvement in yarn quality and processing efficiency (Bobkowicz, 1981).

The development work culminated in the Bobtex Mk 3, which was a single spindle design where units could be bolted together to form a larger machine. Figure 8.8 shows a set-up of ten positions, which was the recommended number per operative. The units were designed to run up to 1000 m/min (with a minimum line speed of 700 m/min).

8.4 The impact of material selection and processing parameters on ICS yarn properties

While much of the data available on ICS emanate from the Bobtex organization, there were a limited number of publications from academics. There were also commentaries in trade journals on the potential and possible limitations of the system.

8.4.1 Initial predictions

Initial discussions centered on the two component yarn system, and the first machine, which was called the 'E7 Bobtex ICSOMAT' (Bobkowicz and Bobkowicz, 1971b), and the yarns produced were called 'BOBYARNS' (Bobkowicz and Bobkowicz, 1970c). The main conclusions from these papers were: the yarn cross-section was variable; the yarn flexibility was improved by using more fibers and less polymer; adding twist to the yarns gave an increase in strength, following a similar trend to the role of twist in other staple spinning systems; and fabrics made from composite yarns had much better abrasion resistance than fabrics from equivalent spun yarns.

The idealized model of the three component ICS yarn is shown in Fig. 8.9 and, as expected, the various components play different roles in the yarn properties. It was argued that Bobtex ICS was the only yarn that could be truly engineered to meet specific properties – through material selection and optimization of processing parameters (Riordon and Bobkowicz, 1974). Table 8.3 is a summary of the impact of material and process on specific yarn properties. In general, the tensile properties were determined primarily by the filament core, whereas the bulk and flexibility were determined by the staple fibers on the yarn surface. Consolidation of the structure was largely governed by the level of false twist that drove the staple fibers into the polymer sheath. Indeed, in extreme cases it was possible to get an 'inverted' structure where the staple component was totally embedded in the sheath, resulting in no surface fibers. It was also acknowledged by Riordon that Bobtex yarns were stiffer than conventional yarns, but it was argued that the difference diminished after fabric finishing, which tended to break some of the secondary mechanical bonds resulting in a softer hand.

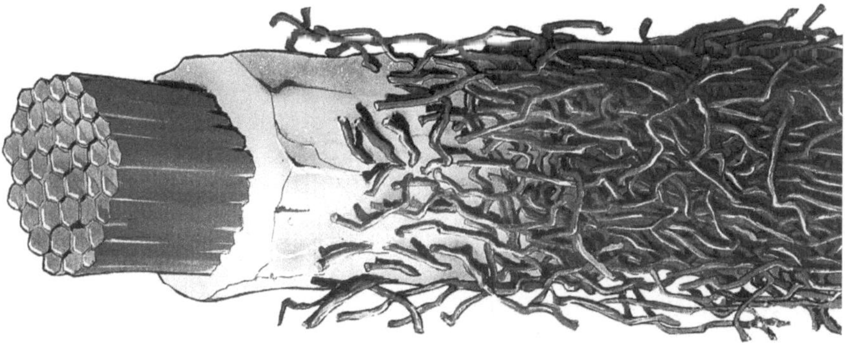

8.9 Idealized yarn structure. Copied from publicity material 'Spin at 1000 m/min' supplied to author by Bobtex International, Leicester, UK.

Table 8.3 Yarn engineering principles

Property	Variable	Comment
Tenacity	Carrier dominated Fiber assisted	Stronger carrier (filament) gives stronger yarn. Longer stronger fibers contribute more
Elongation	Carrier dominated Sheath influenced Process influenced	Extensible carrier give extensible resultant yarn but can be influenced by combined effect of polymer/staple which can restrict the mobility of the filament
Modulus	Carrier dominated Sheath influenced	As with elongation the role of the sheath can modulate the influence of the core – this depends on adhesion of the components
Cover	Fiber percentage Fiber fineness Consolidation	Finer fibers (which means a greater number of fibers for the same percentage) give better cover as does less consolidation (induced by twist)
Bulk	Consolidation Fiber participation Carrier type Polymer density Fiber crimp level	Twist drives the staple into the polymer and gives more consolidation but less bulk. Split film can give bulkier core than filaments and obviously fine denier fibers in the core give more bulk than, say a monofilament
Flexibility	Polymer percentage Carrier type and percentage Twister speed Fiber migration	Twist influences the fiber migration and the ability of the sheath to restrict the mobility of the yarn.
Cost	Components proportions	The use of lower cost components may be possible but yarn quality is essential.

In a subsequent paper, Bobkowicz identified several market areas that he believed offered opportunities for Bobtex yarns, based on their properties and cost. These were broadly classified as 'commodity' and 'new opportunity' yarns, with the former including carpet backing, industrial fabrics, and outerwear, whereas the latter related to high tech applications, such as the use of glass fiber core and the manufacture of flame-retardant products. Other areas of application included fibers not normally associated with traditional textiles, such as short fiber waste, wood pulp, and fibrids, all of which are claimed to process into satisfactory yarns (Bobkowicz, 1975b).

8.4.2 Results from experimental data on yarns

Since the machinery was being developed and improved by the manufacturer and undergoing 'user trials' with selected partners in the industry, there

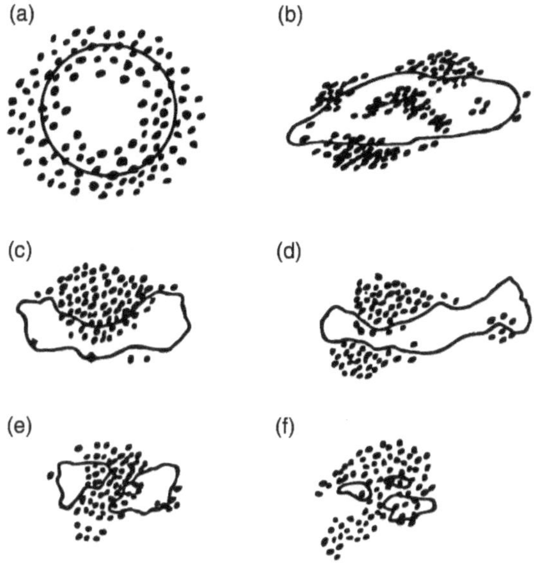

8.10 Schematics of observed yarn structure. Redrawn from Nichols *et al.* (1972).

was some confidentiality about the data generated. This resulted in very limited independent data or critical commentary on the system, and indeed it was apparent that several of the 'independent' reports in trade journals were very largely based on information released by Bobtex Corporation. Mohamed *et al.* carried out studies on both the two component and three component yarns, investigating the structure and selected properties of the yarn and fabric.

His initial investigation (Nichols *et al.*, 1972) focused on a two component yarn, 188 tex made from 55% polyester staple (2.5 dtex, 50 mm) and 45% polyethylene resin, and a plain weave fabric made from the yarns. His findings can be summarized as follows:

- The yarn was bulky and regular but weak and highly extensible (probably associated with undrawn polymer).
- The cross-section was irregular, often ribbon-like, and the staple components were not uniformly distributed, but tended to be concentrated on one side of the polymer. Fig. 8.10 is an illustration of the ideal yarn structure A compared to what was commonly observed B–F.
- In some cases the cross-section indicated that the polymer had split into smaller parts, shown in Figure 9.10 E/F, but this was considered a positive effect, since it could potentially give better fabric cover and softness.

Table 8.4 Physical properties of three component yarns

Percentage composition	Components	Tex	Breaking load (gf)	Elongation (%)	Tenacity (gf/tex)
40/25/35	Cotton (15/16) in./polyester filament (132/32 trevira round)/ polyethylene resin	62.7	770	16.8	12.3
50/25/35	Polyester fortrel (1.5 in. × 1.5 denier)/polyester filament (132/32 trevira round)/polyethylene resin	62.5	924	15.2	14.8
38/21/41	Polyester fortrel (1.5 in. × 1.5 denier)/polyethylene resin	71	917	18	12.9
35/17/48	Polyester fortrel (1.5 in. × 1.5 denier)/fibrillated polypropylene film/polyethylene resin	75	725	16	9.7
27/30/43	Rayon staple (17/16 in. × 1.5 denier)/polypropylene film/ polyethylene resin	184	1866	10	10.1
34/30/36	Polypropylene staple (2 in. × 6 denier)/polypropylene film/ polyethylene resin	192	1490	24	7.8
38/24/38	Nylon staple 2in. × 2.5 denier)/nylon filament (100/32)/nylon resin	45	530	32	11.8
27/23/50	Wool (2 in. 64 s)/nylon filament (100/32)/nylon resin	48	506	40	10.5

Table 8.5 Tensile properties of three component ICS yarn

Yarn	Tex	Breaking load (gf)	Tenacity (gf/tex)	Extension (%)	Initial modulus (gf/tex)	Specific work of rupture (gf/tex/in.)
Yarn 1 Table 8.4	62.7	770	12.3	16.8	243.9	1.5
Yarn 1 with cotton removed	48	470	14.0	27.5	142.8	2.4
Yarn 1 with cotton and resin removed	17.7	543	30.6	20.0	345.9	3.6

- The addition of twist to the yarn was found to increase the strength and extensibility up to a maximum before declining with further twist (exhibiting a similar behavior to that found in ring spun yarns).
- The tensile behavior of the yarn showed three regions: an initial higher modulus region due to staple fibers, followed by a low modulus as the polymer was drawn, and finally a higher modulus as the drawn polymer extended and broke.
- Because of the variation in cross-section and uneven coverage of the yarn, the bending rigidity was found to vary along the yarn length.

In a follow-up paper (Mohamed *et al.*, 1974), the properties of three component ICS yarns were studied. It was claimed that with suitable choice of raw material many different properties could be achieved, and Table 8.4 shows typical examples of yarns. Analysis of cross-sections of the various yarns indicated that the cross-sectional shape was very much dependent on the shape of the core yarn. Thus, while a filament yarn (sample 1 in Table 8.4) yielded a round cross-section, the use of fibrillated tape (sample 4) produced a flattened shape. It was also found that for some yarns there was significant migration of the staple into the polymer (and associated movement of polymer to the surface of the yarn). Yarn sample 1 was evaluated for tensile properties and then re-evaluated after the chemical removal of the staple fibers and the polymer. Table 8.5 shows that, while the strength of the yarn was higher when the staple fibers were present, the tenacity was lower than the filaments. Additionally, the presence of polymer and staple reduces the elongation of the yarn.

8.4.3 Results from experimental data focusing on fabrics

Nichols *et al.* (1972) made observations on the cross-section of plain fabrics that had been woven from two component yarn, and there was some concern that there may be shedding of the staple component during weaving. There had also been concern about the stiffness of the yarns, since this could result in harsh handling fabrics, which would severely curtail their potential areas of application. Lord (1974) carried out some experimentation on denim manufactured from blue cotton warp and Bobtex weft yarn. The weft was 45 tex made from 35% (38 mm, 1.3 dtex) rayon, 35% filament nylon carrier, and 30% polyethylene polymer. It was expected that washing would soften the fabric; however, there was little change after several washing and drying cycles. This result is thought to be associated with the combined effects of the washing, breaking the bonds, and softening the fabric, which was countered by fabric shrinkage, which made the fabric stiffer. In strength, tear, and abrasion tests the fabric performance was very good when compared to a fabric made from 100% cotton, and in some cases the fabric properties improved after washing.

8.5 Commercialization of the process

It is clearly evident that Bobtex managed to maintain active press coverage of its developments, and particularly the performance benefits of its yarn and the economic advantages of its process. In an article on cost issues, Lord (1973) indicated that there may be potential disadvantages in moving from a two component to a three component yarn. He explained that the core yarn (carrier), which must be greater than 10% of the total yarn weight, represented the most significant cost, although the use of fibrillated tape could offer some savings. For stronger yarns the core component should be increased, but it was essential to ensure that there was always adequate staple fiber to give good cover. He also indicated that there were many available resins that could be used, and it may have thus been possible to use lower cost resins in less critical applications and potentially reduce the overall raw material costs. Typically, fine yarns (<30 tex) required about 30% resin, whereas coarser yarns (>100 tex) could be made with 20%. There were many staple fibers that could be used, but finer fibers, while being generally more expensive, gave better cover and softness to the yarn.

Through various trade shows and conferences, the textile industry was kept aware of Bobtex technology. Frequent press releases before and during the international Textile Machinery Exhibition in Milan in 1975, indicated that Mk 1 Model B, an updated version of the machine, would be exhibited. It was also announced that Leesona Corporation had been granted a non-exclusive license to commercialize the Bobtex ICS system. It was noteworthy that Johnson (1975) produced a short, technical article indicating the significance of tension on the carrier and particularly the problem of maintaining constant tension when processing at high speeds. Specially designed

Table 8.6 Examples of applications of ICS yarn

Application	Remarks
Produce packaging	Woven or knit bags, strong, lightweight, spun surface does not damage produce, waste fiber used for cover.
Industrial and environmental	High abrasion resistance, unique constructions for chemical and surface needs.
Decorative, household	Easy multi-color, high wear.
Fashion outerwear	Superior knittability, Shetland colors easy, wool and naturals desired hand and softness, fancy knits.
Protective	Anti-flam construction inherent: clothing, drapery, upholstery.

hysteresis tensioners were incorporated into the threadline to ensure constant tension.

To stimulate machinery sales, the economic advantages of Bobtex ICS were promoted, and this included the encouragement of the use of polypropylene as the polymer of choice (Bobkowicz, 1975c). In a subsequent paper, Bobkowicz and Slanik (1978) further indicated that alloys of polypropylene could be used as polymer and, while this was relatively inexpensive, it offered significant benefits in terms of reduced yarn stiffness coupled with better bonding of staple fiber components. In this paper it is also shown that for coarse count yarns, payback on investment for a 20 position ICS installation (from bale to yarn) could be less than one year when the machinery was operated at the highest production speeds. Typical areas of application for the yarns are given in Table 8.6. Similar reports concerning the benefits of the new alloy polypropylene appeared elsewhere, along with indications of the possibilities of producing economic flame-retardant fabrics (Anon, 1978).

While Bobtex was promoting the ICS system, there was also a lot of news coverage surrounding a development in the UK, which was the first commercial production unit of Bobtex ICS (Anon, 1981a). Remora Textiles Ltd., led by Alan Barnes, had moved to a purpose-built building in Leicester, which was the heart of the UK knitting industry, and set up a group of ten Bobtex ICS machines plus the necessary preparatory machinery. Barnes had purchased a group of machines that had been used in different processing mills on a trial basis, and assembled them into a single production unit (Lennox-Kerr, 1980a). The report was very favorable and, by working with the local knitting industry, it appeared that Remora might make a success of Bobtex. One particular comment was that the rigidity of the core structure actually presented benefits in certain applications, such as pleated knitted skirts, which were prone to sagging when made from staple yarns.

Other reports (Anon, 1980b) indicated that Remora had conducted market research and focused production of yarns using wool, acrylic, and wool/acrylic blends, and making yarns of 65, 80, and 104 tex with 167, 220, and 330 dtex polyester core yarns. The yarns were claimed (Anon, 1980a) to offer better stitch clarity, and it was argued that the claimed superior abrasion resistance would be readily applicable to the sock market. Lennox-Kerr (1980b) reported that as well as offering the advantage of 10% greater bulk, the yarns were also typically 10% cheaper than their traditionally spun equivalents. It is interesting to note that in the initial stages Remora worked closely with fiber producers Monsanto and Viscosuisse to optimize fiber and machine parameters for the core and staple components of the yarns (Anon, 1981b).

Early in 1981, Remora was spinning 3.5 tons per week on ten machines, and two additional machines were due for delivery. The goal was to add an additional machine to give a capacity of 20 tons per week (Anon, 1981b). Unfortunately, the dreams of the various reporters were never realized, and Remora and its machines disappeared. The reasons for this failure are complex, since the company was exploring alternative markets to develop the business, including more technical yarns for flame resist products (Barnes, 1984). Unfortunately, this was a time when the UK textile industry was downsizing and global completion was becoming a reality. International sourcing meant that products were being imported for prices cheaper than they could be manufactured in the UK, largely because of labor costs, and thus the market for yarns for the commodity market was waning.

8.6 Assessing the advantages and disadvantages of ICS technology

Bobtex ICS was part of the boom that occurred in new spinning methods in the 1970/1980 era, and to yarn technologists it seemed to offer a lot of potential, but it perhaps suffered from the mindset 'if it's strong and I can spin it fast and economically then it will succeed.' Bobtex was only one of many systems that either died or ended up in small niche markets despite offering significant promise.

There are several aspects which must be considered when determining the 'quality' of a yarn:

- Performance parameters
  - Physical attributes, which are normally measured in a testing laboratory, such as strength, uniformity, bulk, abrasion resistance, etc. These can be readily measured but the translation of these properties through to fabric and end-product must also be evaluated.

  - Processing performance, including some measure of how well the yarn spins, knits, and weaves. This should also take into consideration how easily the fabric can be finished and ultimately converted into a product (is it easy to sew?)
- Cost
  - This should consider the cost to manufacture the yarn, through to the final product including special finishing treatments that may need to be used.
- Aesthetics
  - The majority of apparel textiles are bought based on perceptions of aesthetics, and this includes appearance, color, feel, and these so-called 'tactivisual properties' (i.e. those perceived through touch and sight) are based on personal preferences.

In order to be successful, a yarn must be adequate in all of the above areas, and not just outstanding in one or two.

Despite the many articles to the contrary, there were certainly problems with Bobtex as an apparel yarn. While the use of different polymers to bind the staple to the core resulted in some improvement in aesthetics, the yarns and fabrics were still distinctly different from the products they were trying to replace, and it is likely that the apparel market should not have been considered a real opportunity for Bobtex yarns.

8.7 Future trends

It is interesting to note that, with the growth in demand for composite products, there is a renewed interest in composite yarns. One particular area of focus is the use of natural fibers in a synthetic resin; however, there is an issue in obtaining adequate distribution of the resin through the fibers. A possible solution to this problem is to use composite yarns, where the natural staple fiber has been combined with a filament, which can be formed into fabric and then molded. When heat is applied, the filament melts and provides intimate penetration of the staple component (Jing, 2012). A potential advantage of this approach is that the orientation of the staple fibers can potentially be controlled by the technique used to produce the composite yarn (since this is significantly influenced by the degree of alignment of the staple component).

A further potential area of application of composite yarns is in the processing of much shorter fibers, such as for example carbon nanotubes. It is interesting to note that one of the many patents filed by Bobkowicz pertained specifically to the processing of fibrids (specified as 1/8–1/2 inch and 0.2 and 1 denier), and it was argued that there was no other satisfactory technique of spinning these fibers into yarns. (Bobkowicz and Bobkowicz, 1971a). This technology was based around the two component system in which the fibrids were embedded into a molten polymer, which may be drawn during yarn formation. The patent claims: 'Thus, according to the invention there is first of all provided a novel composite yarn of high absorbency, high retention of absorbed liquid, outstanding durability, excellent tensile strength, superior stretch and recovery characteristics, easy performance in washing and dry cleaning, good heat sealability, excellent surface softness as well as fibrous touch and, with selected fibrids, a good chemical resistance and dielectric strength' (Bobkowicz and Bobkowicz, 1971a).

The idea of using a polymeric adhesive to bind very short fibers to a core yarn was also proposed by Schutte (1976), where the 'fibers' were flock, from wood pulp, with lengths of less than 0.25 in. While this patent shows idealized structures and provides examples of yarn properties, there is no clear indication of the technology used to manufacture the yarns. While flocked yarns, using very short fibers to cover the surface, tend to rely on different technology from the original ICS concept, which was based around an extrusion process, the product from flocking is somewhat similar. A further major difference between flocking and the original ICS is that the fibers are added to the core, which is covered with an adhesive, by means of electrostatic forces (Haranoya and Iwamoyo, 1989). There is presently considerable interest in yarn flocking technology, which is claimed to be driven by market forces, and there is active research into better production techniques.

8.8 Conclusion

While the Bobtex ICS system was a sophisticated system and offered the opportunity to engineer certain yarn characteristics, it could never offer soft 'aesthetically pleasing' yarns. Composites are fundamentally stiffer, and this is evidenced in nature where fibers in a matrix create more rigid structures. Indeed, this problem was experienced in other spinning systems, where core fibers were bound together, since it prevents relative movement of fibers and thus the product will always be relatively stiff. There never seemed to be market research aimed at identifying a market where a stiff yarn could be advantageous. Unfortunately, the quest seemed to be commodity yarns, and this was a major mistake in developing a new market for a new yarn.

Even Remora, who were hailed as the future of the UK yarn manufacturing industry, floundered by selecting the wrong market. While it seemed likely that here could be a market for wool and wool/acrylic blend yarns, the wool market is probably the most difficult to introduce change, with purchasers of wool garments being typically conservative and not ready to accept changes. Added to this problem is the fact that wool is a fairly stiff fiber and as such is not the ideal choice for inclusion in a yarn that is further stiffened by the structure. While Remora was considering technical applications, they unfortunately spent considerably more effort developing product for the apparel industry, which proved to be an expensive exercise with limited opportunity.

It is evident that this approach of misidentifying trends was a problem at the beginning of the development of Bobtex, when for example the yarns were proposed as candidates for inclusion in denim fabrics. A major fashion feature of denim is that it fades and can be distressed; however, claims made for Bobtex were that its wear characteristics, including abrasion resistance, were much better than other yarns. This would appear to restrict the application of Bobtex to denim that would be used only for work-wear.

A final question would be: 'if Bobtex were to be introduced today, would it be successful?' If the strategy were 'to create yarns to meet technical requirements, rather than creating a yarn and then trying to find a use for it in an apparel application,' then it is believed that it could be successful in specialist, but high value, markets such as medical, technical, and sports applications. As indicated above, there is a growing demand for yarns made from 'non-textile' fibers and, while there are coating techniques available, these tend to lack the elegance and flexibility of Bobtex.

As a long time 'yarn technologist,' if given the opportunity to acquire one piece of spinning equipment, a Bobtex ICS machine would be always the first choice, and this is a view shared by colleagues with similar interests.

8.9 References

Anon (1978), Update on unconventional spinning, *Textile Industries*, March.

Anon (1980a), Leicester's revolutionary new spinning mill, *Knitting International*, August.

Anon (1980b), Rethink yarn manufacture, *African Textiles*, November, 13–15.

Anon (1981a), British plant makes a go of the Bobtex yarn system, *Canadian Textile Journal*, January, 44.

Anon (1981b), Leicester spinners are the world leader in Bobtex technology, *Textile Month*, February, 38–39.

Barnes A. (1984), Flame resistant yarns and fabrics, and furnishings incorporating such fabrics, EP0106840.

Bobkowicz E. and Bobkowicz A.J. (1967), Textured filament yarns, USP 3347727.

Bobkowicz E. and Bobkowicz A.J. (1970a), Universal open-end spinning method of multicomponent yarns production, USP 3494118.

Bobkowicz E. and Bobkowicz A.J. (1970b), Method and apparatus for ringless spinning of fiber-polymer yarns, USP 3548581.

Bobkowicz E. and Bobkowicz A.J. (1970c), The Bobtex ICS process, *Canadian Textile Journal*, April, 77–84.

Bobkowicz E. and Bobkowicz A.J. (1971a), Composite yarns and method of manufacture, USP 3621646.

Bobkowicz E. and Bobkowicz A.J. (1971b), The Bobtex yarn-making technology: its future and implications, *Textile Research Journal*, **41**, 773–778

Bobkowicz E. and Bobkowicz A.J. (1974a), Composite spinning, USP 3844097.

Bobkowicz E. and Bobkowicz A.J. (1974b), Aerodynamic spinning of composite yarns, USP 3848403.

Bobkowicz E. and Bobkowicz A.J. (1975a), Composite yarn forming method and apparatus, USP 3924396.

Bobkowicz A.J. (1975b), Bobtex integrated composite spinning yarns: applications and possible offsprings, *Modern Textiles*, June.

Bobkowicz A.J. (1975c), Polypropylene and ICS processing, *International Conference on 'Polypropylene Fibers in Textiles'*, University of York, UK, 1 October, paper G7.

Bobkowicz E. (1976), Method and apparatus for the manufacture of composite yarn products, USP 3987612.

Bobkowicz A.J. and Slanik J. (1978), Bobtex integrated composite spinning: new use of fibers and polymer alloys to modify filaments' surface, *Canadian Textile Journal*, May.

Bobkowicz A.J. (1981), POY yarn compositions, USP 4244174.

Denton M.J. and Daniels P.N. (2002), *Textile Terms and Definitions*, Manchester, UK, The Textile Institute.

Haranoya T. and Iwamoyo S. (1989), Flocked yarn and method for manufacture, USP 4886693.

Jeddi A.A.A., Johari M.S. and Merati A.A. (2009), A study of the structure and physical properties of cotton-covered nylon filament core-spun yarns, *Journal of the Textile Institute*, **88**, 12–20.

Jiang J. and Chen N. (2012), Preforms and composites manufactured by novel flax/polypropylene co-wrap spinning method, *Journal of Composite Materials*, Available from http://jcm.sagepub.com/content/early/2012/01/18/0021998311430155 (Accessed 27 April 2012).

Johnson F.M. (1975), Bobtex ICS machine design considerations, *Canadian Textile Journal*, October.

Jou G.T., East G.C., Lawrence C.A. and Oxenham W. (1996), The physical properties of composite yarns produced by an electrostatic-filament charging method, *Journal of the Textile Institute*, **87**, 78–97.

Lennox-Kerr P. (1980a), A yarn for the times?, *Textile Asia*, November, 11.

Lennox-Kerr P. (1980b), Bobtex: the system takes off at Remora Textiles, *Textile Institute and Industry*, November, 307–308.

Lord P.R. (1973), Three-component Bobtex yarns raise cost question, *Textile Month*, September, 85–89

Lord P.R. (1974), Bobtex denim: some surprises, *Textile Industries*, May.

Mohamed M.H. (1973), Bobtex (ICS) new yarn and new technology, *Canadian Textile Journal*, September, 91–94.

Mohamed M.H., Rochow T.G. and Heeralal R. (1974), Some structural and physical properties of yarn made on the integrated composite spinning system. Part II: Three-component yarn, *Textile Research Journal*, **44**, 206–213.

Nichols L.D., Mohamed M.H. and Rochow T.G. (1972), Some structural and physical properties of yarn made on the integrated composite spinning system. Part I: Two-component yarns, *Textile Research Journal*, **42**, 338–344.

Nield R. and Ali R.A. (1977), Open-end-spun core-spun yarns, *Journal of the Textile Institute*, **68**, 223–229.
Orleck H.G. and Ulku S. (2007), Effects of spandex and yarn counts on the properties of elastic core-spun yarns produced on Murata Vortex spinner, *Textile Research Journal*, **77**, 432–436.
Riordon P.R. and Bobkowicz A.J. (1974), Bobtex ICS as a new system for making yarns, *Textile Manufacturer*, July, 26–30.
Selling H.J. (1971), *Twistless Yarns*, Watford, Merrow Publishing Co. Ltd.
Schutte R.W. (1976), Flocked elementary element and structures made therefrom, USP 3 968 283.
Xie Y., Oxenham W. and Grosberg P. (1986), A study of the strength of wrapped yarns. Part I: The theoretical model, *Journal of the Textile Institute*, **77**, 295–304.

9
Wet spinning of synthetic polymer fibers

B. OZIPEK and H. KARAKAS,
Istanbul Technical University, Turkey

DOI: 10.1533/9780857099174.2.174

Abstract: This chapter deals with the production of wet-spun man-made fibers. In this context, major commercially available wet-spun fibers are discussed. The principles of wet spinning, including extrusion, coagulation, and drawing stages are described. Parameters affecting polymer solution and spinning process which, in turn, determine the resultant fiber behavior and performance, are briefly explained. Special attention is given to acrylic and cuprammonium fiber production and resultant fiber characteristics. End use properties of wet-spun fibers are also evaluated. Finally, new developments in the wet-spun fiber production and future prospects are outlined.

Key words: wet spinning, extrusion, coagulation, stretching, fiber properties.

9.1 Introduction

Solution spinning techniques require dissolving of polymer in a solvent without degrading the polymer. The wet spinning process involves filament extrusion into a nonsolvent. Since it is a gentle process, applying lower temperatures, it has been the preferred method for production of fibers that cannot be melt spun. The wet spinning process offers the advantage of producing a wide variety of fiber cross-sectional shapes and sizes. Higher production speeds have been successfully attained in wet spinning over the last few years. This chapter focuses on the principles of wet spinning, with special emphasis on cuprammonium rayon and acrylic production.

9.2 Principles of wet spinning

Wet spinning of fibers requires pumping of the polymer solution through the fine orifices of a spinneret into a coagulating bath, and drawing off as continuous filaments by means of take-up rollers. The bath removes the solvent from the as-spun filaments so that they become solidified. They are collected together to form a continuous tow or rope.

The wet spinning method was first used in the manufacture of rayon (Chardonnet Silk) by extruding the alcoholic solution of the cellulose nitrate through a mouthpiece dipped in cold water. As the stream of collodion (solution of cellulose nitrate) passes through the water, the filaments are stretched and solidified. Solidification of the filament takes place from surface to the core. Originally, the viscose solution, which is ripped, is forced by compressed air to the spinning frame. The pump used for production of viscose is usually of the gear type through which the viscose passes to the filter, and subsequently through a spinneret. The spinnerets are generally submerged in an acid bath, and are usually arranged so that filaments emerge almost at right angles to the surface of the bath. As the viscose solution passes through the jets into the acid bath, it solidifies into filaments, owing to regeneration of the cellulose. Significantly lower speeds are used in wet spinning compared to melt and dry spinning techniques, as the spinning takes place using fluids of higher viscosity that generate high tension on the filaments. This problem can be overcome by using low speeds.

Wet spinning of *acrylic fibers* may be carried out in a variety of precipitating baths. Glycerol or water is used in the baths. Meanwhile, no specific method has been released for a particular yarn production yet. Differences in bath composition result in significant differences in the properties of the acrylic fibers produced.

Some of the baths used for acrylic fibers can be summarized as follows:

- A 20% solution of polymer in ethylene carbonate is heated to 120°C, and extruded into a bath of 80% dipropylene glycol and 20% ethylene carbonate at 130°C. The yarn is washed with water at 80°C, stretched ten-fold at 150°C, and finally relaxed at 140°C (Fig. 9.1).
- A 20% solution of polymer in dimethyl acetamide is spun into glycerol at 140°C and wound on to a bobbin, which is washed free of glycerol with water and dried. The yarn is maintained under tension during its passage through the glycerol bath. Sometimes two glycerol baths may be used, the first at 120°C and the second at 170°C.
- A 20% solution of polymer in dimethyl acetamide is spun into a 40% solution of calcium chloride in water at 90°C.
- An 18% solution of polymer in dimethyl acetamide is spun into a mixture of 2 parts dimethyl acetamide and 1 part water. As the yarn emerges from the bath, it is washed with water, which flows counter-currently into the spinning bath at such a rate that the bath composition is kept constant. Dimethyl acetamide is continually being added to the polymer while it is extruded.
- In the case of Courtelle®, which is a commercial type of acrylic fiber by Acordis Acrylic Fiber and formerly Courtaulds, the polymer was originally dissolved in a strong aqueous solution of calcium thiocyanate and

9.1 Production of wet-spun acrylic fiber. (*Source*: Courtesy of Aksa, a member of AKKOK Group of Companies, Turkey.)

9.2 Extrusion of acrylic fibers from the spinneret. (*Source*: Courtesy of Aksa, a member of AKKOK Group of Companies, Turkey.)

was spun in an aqueous bath. Besides dimethyl acetamide, dimethyl formamide, 60% zinc chloride, and 70% nitric acid are solvents that can be used for the preparation of acrylic filaments (Fig. 9.2).

Alginate fibers are also produced by the wet spinning method. The alginic acid is reacted in the solid state with sodium carbonate and the resulting sodium alginate is used for spinning. The grade used for fiber spinning has a viscosity of about 60–80 centistokes in 1% solution at 25°C; lower grades (lower viscosities) give poor fibers. Much higher grades are more difficult to dissolve and filter, but may yield a better quality fiber than any made from seaweed. A solution of 8–9% sodium alginate is prepared and is sterilized by the addition of a bactericide. It is filtered and spun into a coagulating bath containing normal calcium chloride solution, 0.02 N hydrochloric acid solution, and a small quantity of a cationic surface active agent. As the sodium alginate is ejected from the jet it is precipitated in the form of calcium alginate filaments. The filaments are drawn together, washed, lubricated, dried, and wound. The function of the hydrochloric acid is to prevent closure of the jets by the growth of calcium alginate on their sides; that of the cationic agent is to prevent filament adhesion.

Aramid fibers are produced via the wet spinning process. Kevlar and Twaron are both p-phenyleneterephthalamides (PPTA), the simplest form of the AABB para-polyaramide. PPTA is a product of p-phenylenediamine (PPD) and terephthaloyl dichloride (TDC or TCl). Production of PPTA relies on a co-solvent with an ionic component (calcium chloride, $CaCl_2$) to occupy the hydrogen bonds of the amide groups, and an organic component (N-methyl pyrrolidone (NMP)) to dissolve the aromatic polymer. After production of the polymer, the aramid fiber is produced by spinning the dissolved polymer from a liquid chemical blend to form a solid fiber. The polymer solvent for spinning PPTA is generally 100% anhydrous sulfuric acid (H_2SO_4).

Dynel, a *modacrylic fiber*, is a wet-spun fiber. Two chemicals, i.e. acrylonitrile and vinyl chloride, are co-polymerized in the ratio of 2:3 and the copolymer is dissolved to a 21% solution in acetone. The acetone solution is de-aerated, filtered, and then extruded by the usual wet spinning technique into a water bath, where it coagulates to form continuous strands of filaments.

9.3 Types of fibers used

Production of viscose rayon and cuprammonium rayon will be described as they are among the oldest wet-spun man-made fibers.

9.3.1 Production of viscose rayon

Viscose rayon fiber production benefits from the wet spinning process, as it necessitates spinning of filament from a liquid of regenerated cellulose in a spinning bath. The viscose process consists of the following steps (Fig. 9.3):

1. Steeping: cellulose pulps in large white sheets are immersed in vats containing 17–20% aqueous sodium hydroxide (NaOH) at a temperature

ranging between 18°C and 25°C. The cellulose sheets are swollen, and cellulose is converted into alkali cellulose.

2. Pressing: the swollen alkali cellulose sheets are pressed so that a press weight ratio of 2.7–3.0 is obtained. This is the ratio of the weight of pressed alkali cellulose sheet to air-dried pulp.[1]
3. Shredding: the sheets of alkali cellulose are mechanically shredded in the form of 'crumbs,' which are finely divided particles. This process provides increased surface area to increase the reaction ability of alkali cellulose.
4. Ageing: ageing aims to decrease the polymerization degree of alkali cellulose to the desired level. This process allows the alkali cellulose under controlled humidity and temperature conditions to obtain the required viscosity for spinning.
5. Xanthation: this step is the reaction of aged alkali cellulose with carbon disulfide to have a resultant mass of cellulose xanthate. The cellulose crumbs become orange in color at the end of this process.
6. Preparation of the viscose solution: the cellulose xanthate is mixed with an aqueous caustic solution in a stirring tank to form the viscous solution. It is called 'viscose' as this solution has a high viscosity. The viscose solution is thoroughly filtered to remove any undissolved material, and then de-aerated for removal of all air bubbles.
7. Ripening: in order to decrease the viscosity of the prepared solution, it is allowed to ripen for a period of time. During this process, the xanthate groups revert to cellulosic hydroxyls and free CS_2.
8. Filtering and degassing: the solution is then filtered once more to prevent the disruption of the spinning process. The air bubbles are removed, as otherwise they may lead to breaks or weak spots during spinning.
9. Spinning: Viscous rayon filaments are spun by wet spinning process. The viscous fluid is delivered to the spinneret, which is immersed in a spin bath containing sulfuric acid, sodium sulfate and zinc sulfate. As the rayon filaments rapidly coagulate they are stretched, and cellulose xanthate decomposes to regenerated cellulose.
10. Drawing, washing and cutting: the viscose rayon filaments are drawn to exhibit orientation of molecular chains along the fiber axis. This step is intended to obtain the required tenacity for use as textile fibers. The impurities and salts are washed away. This is followed by cutting if the filaments are to be converted to staple form.

Instead of batchwise processes, continuous processes are predominantly used in the viscose process. Process continuity has been guaranteed via improved level of automation.[1]

The environmental impact of viscose processes, energy consumption, and use of limited sources and disposability of the products have been major

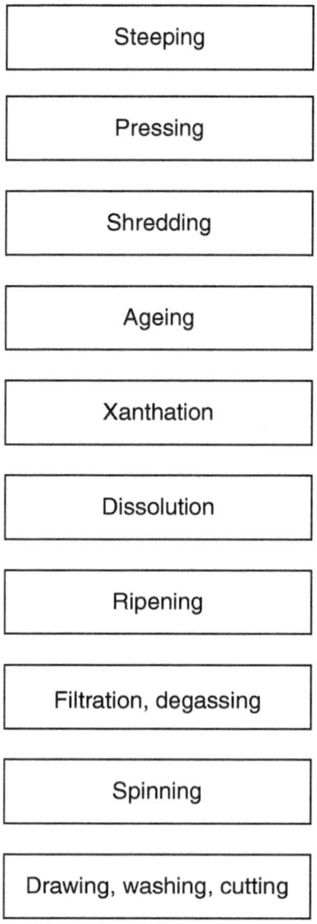

9.3 Viscose fiber production steps.

issues in the recent years. Additionally, viscose fiber manufacturing plants have effluent generation, as viscose fiber manufacturing can be regarded as a wet process using high amounts of process water. Use of carbon disulfide is an important burden for the environment. Sulfur recovery cannot be maintained at 100%, as it forms sulfur compounds such as CS_2, H_2S, COS, and SO_2. These compounds can be found in exhaust gases, spin baths, and wastes.[2] Effective recovery of CS_2 and H_2S, and attempts to reduce the use of chemicals such as NaOH and H_2SO_4, have been the most important objectives of the viscose industry's environmental concerns. CS_2 is an expensive material, and therefore its reuse is preferred. The high amounts of CS_2 and H_2S are removed by degassing for easier recovery. Spinning liquors are evaporated and CS_2 is mostly recovered by condensation or adsorption from high and low concentrated process exhaust gases respectively. In most viscose plants, precipitated crystals of Na_2SO_4 (Glauber's salt) are mostly heated and converted into anhydrous sodium sulfate for use. The viscose spinning bath contains zinc sulfate, which has also to be recovered, due to environmental concerns. Effluent treatment is made through chemical or biological treatment.[3]

9.3.2 Production of cuprammonium rayon

The cuprammonium rayon process was first used in America by the American Bemberg Corporation in 1926, in a plant located in Elizabethton, Tennessee. The trade name of this rayon yarn was Bemberg. In fact, the technology for the production of Bemberg was first invented by a German, J.P. Bemberg, and the first commercial production began in Germany in 1897. The parent company J. P. Bemberg was the German affiliate of VereinigteGlanzstoffFabriken. Today, Asahi Chemical Industries Corporation in Japan and Bemberg SpA in Italy produce Bemberg.

Cotton linters, the short fibers on the surface of cotton seed after ginning, are used for this process, and the solvent used here is a mixture of copper hydroxide and aqueous ammonia. The solution of copper hydroxide (or basic copper sulfate), aqueous ammonia, and cellulose is filtered and degassed, followed by the addition of a small amount of aqueous sodium hydroxide. A process called 'stretch spinning,' invented by Edmund Thiele in 1901 in Germany and first used by J. P. Bemberg AG, was used for the production of Bemberg. The filaments are formed during the passage of the solvent from the spinnerets through a cone-shaped tunnel with degassed and deionized water. Ammonia and some of the copper are removed during the passage of the filaments. Coagulation takes place in this tunnel, and meanwhile the coagulated dope is elongated hundreds of times, resulting in fiber formation. The yarn is then washed thoroughly with water and then dried.

Asahi Chemicals Industries Corporation developed the double or triple funnel system and a bath between 1950 and 1990. A number of patents were taken to improve the 'two-step funnel' concept. In 1962, Aizawa *et al.* invented an improved method for the stretch-spinning process by discharging the spinning solution into a stream of coagulating liquid flow. The sectional view of the invention and the conventional funnel type can be seen in Fig. 9.4. The apparatus consisted of a conical tube, a long straight tube, and a coagulation zone. The cuprammonium spinning solution with a coagulating liquid was extruded into a stretch coagulation zone and stretched. Then they were passed into a further coagulation zone, prior to completion of the first coagulation process, where the flow rate of the coagulating liquid was diminished. Coagulation was completed in a relaxed state. Thereby, cuprammonium rayons of superior tenacity, homogeneous sectional structure and easy resin finishing character were achieved.[4–7]

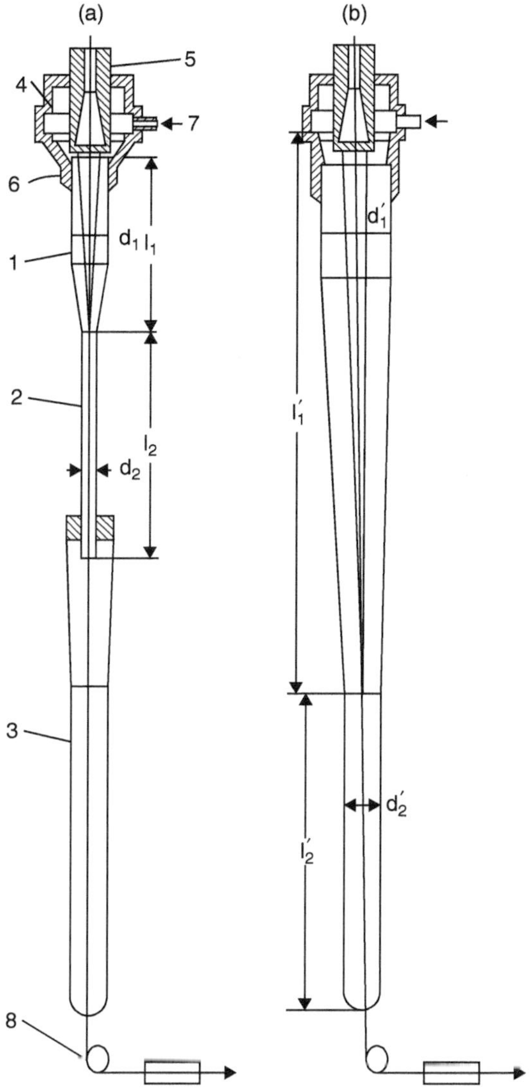

9.4 (a) Sectional view of the apparatus invented by Aizawa *et al.*, (b) Typical conventional funnel type apparatus.[7]

9.3.3 Production of lyocell

Lyocell fibers were first developed and manufactured as Tencel™ fiber by Courtaulds Fibers, UK, in the 1980s. Lenzing AG, Austria produces lyocell under the brand name 'Lyocell by Lenzing.' The generic name being 'lyocell,' the fibers are produced under brand names that include Tencel (Acordis, previously Courtaulds), Lenzing Lyocell (Lenzing), Newcell (Akzo Nobel), and special brands including Seacell (Zimmer AG).[2]

Lyocell and viscose are produced from wood pulp but with different manufacturing processes. Lyocell is produced by direct dissolving process and involves no derivative, unlike viscose rayon.

Solution of hot N-Methylmorpholine-N-oxide (NMMO), a cyclic amine oxide that has proved to be an excellent solvent for cellulose, is used to dissolve wood pulp. The viscous solution, after filtering, is spun into a coagulation bath of aqueous solution of the amine oxide. (Gupta, Kothari). The spinning process is followed by washing and drying of the filaments. The process gives the advantages of a reduced number of processing stages, almost 99% solvent recovery, and fiber properties unattainable by the classic production processes.[8–10] Only very small amounts of NMMO remain in the waste water, which is further treated biologically.

Lyocell fibers offer advantages with their properties and environment friendly production processes. Fibrillation, which is among the most important properties of lyocell fiber, can be used to produce an appealing fabric touch called 'peach-skin effect'.[2]

9.4 Fiber properties

Fiber properties are greatly affected from the process parameters of the wet spinning process, which is a gentle process on the fibers. The process also has the advantage of offering a wide range of cross-sections.

9.4.1 Cross-section

Wet spinning offers a wide range of cross-sectional shapes, including circular and bean-shaped. Wet-spun fibers have also been highly stretched approximately to round. This applies to acrylics, cuprammonium, casein, lyocell, alginate, or so on. It is worthwhile to note that this is mainly because of the usage of aqueous bath in wet spinning. Lyocell fibers have a circular cross-section and smooth fiber surface.

9.4.2 Density

The density of acrylics is around 1.17 g/c.c. at 25°C. For viscose fibers, it is 1.52 g/c.c., which is higher than cellulose acetate but equivalent to cuprammonium. Casein fibers have a density of 1.29 g/c.c., very similar to wool. On the other hand, alginate fibers have a high metal content; calcium alginate contains about 10% calcium, and this brings about the high specific density of 1.75 g/c.c. Lyocell also has a comparatively high density of 1.5 g/c.c.

9.4.3 Mechanical properties

The tenacity of casein fibers is about 0.8–1.0 g/den, and the elongation at break is about 15%. When wet, the tenacity is very low; this is the main defect of casein fibers, but the elongation at break is correspondingly higher.

Ordinary viscose fiber is reasonably strong, with tenacity about 2.6 g/den. This compares very favorably with cellulose acetate, which has a dry tenacity of 1.3–1.7 g/den. The wet strength of viscose fiber was originally about 1.4 g/den. A great deal of work has been carried out to increase the wet strength of viscose fibers, and today its different types have rather good wet strengths.

Calcium alginate fibers have a dry strength comparable with that of viscose fibers, but their wet strength is low; their extensibility is sufficiently high to meet most textile requirements.

Lyocell is the strongest cellulosic fiber when dry, even stronger than cotton or linen, and is stronger than cotton when wet. Lyocell is much stronger than viscose fiber when wet. This property of high wet strength usually determines the extent to which fabrics can be machine washed successfully. Lyocell fibers have a high degree of orientation and crystallinity.

In the case of acrylics, the dry tensile strength is 2.5 g/den and extension at break is 35%; corresponding figures for wet fiber is 2.0 g/den and 44%. In a study by Aghanouri *et al.*, the tenacities of dry-spun and wet-spun acrylic fibers have been compared after photooxidative degradation, as photooxidation of acrylic fibers can lead to discoloration and reduction in mechanical properties. It was found that degradation was dependent on the spinning method. Wet-spun fibers had a greater decrease in tenacity compared to dry-spun fibers.[11]

9.4.4 Moisture regain

Viscose rayon has a moisture regain of 11–13%, whereas lyocell has 11.5% at standard conditions (20°C, 65% relative humidity). Acrylic's moisture regain value ranges between 1 and 2%. Cellulosic fibers have polar groups and show high regain, whereas nonpolar fibers have no or little regain.[12]

9.5 Factors affecting wet spinning

The spinning speeds in wet spinning are lower compared to dry spinning, which has even lower speeds than melt spinning as the process involves filament formation by coagulating in a spin bath. Application of low temperatures and the flexibility of having all spinning and post-spinning operations in a continuous process compensate for the low spinning speeds, and make this method favorable.

The wet spinning process consists of preparation of the polymer solution, extrusion of polymer solution into a liquid bath, coagulation in this bath and further processes applied to the spun fibers. Polymer concentration in the dope depends mainly on polymer solubility and spinning pressure limitations.[1,13] The fiber formation in the coagulation bath is a complex process and it involves parameters of bath composition, temperature, extrusion and take-up velocity.

Mass transfer rate difference is among the most important factors in the coagulation of the outer layer of the dope, and is determinative in the resultant properties and microstructure of the as-spun fibers.

Mass transfer in wet spinning has been studied extensively in the literature. Rende built a physiochemical model to understand the relationship between coagulation of polyacrylonitrile and the coagulation bath variables. The model shows that polymer is precipitated at a distance from the fiber axis that varies with time.[14]

Ji investigated the effect of coagulation variables and defined the mass transfer rate difference between solvent and coagulant. Polyacrylonitrile and dimethylsulfoxide (DMSO) dopes were immersed in coagulation baths with different coagulants and weight loss was observed in the dope during the coagulation process.[15] Liu *et al.* have investigated the effect of variables on the coagulation of cellulose from various solutions. The coagulation rate was determined to be a function of molecular volumes of coagulant, bath temperatures, bath compositions, and cellulose concentrations.[16]

9.6 Applications and future trends

New wet spinning methods are being developed and used for various fiber formations. Hollow tubular chitosan fibers have been produced by the wet spinning method, which involves successive washing with water to stop the coagulation process. The chitosan dope is obtained by dissolving chitosan with acetic acid, while the coagulation bath is an aqueous solution of sodium hydroxide. Interrupted wet spinning allowed obtaining hollow chitosan fibers as the reaction in the coagulation bath was interrupted by water washing baths. This process allows the forming of a chitosan outer membrane gel and a core chitosan solution, followed by removal of the core chitosan solution and formation of the hollow fiber structure. Use of dilute chitosan dope lower than 2.5% (w/w) led to a smaller internal fiber diameter of less than 200 μm.[17]

Wet spinning methods are also being explored by fiber scientists to develop conducting polymer fibers. Continuous poly(3,4-ethylenedioxythiophene):poly(styrenesulfonate) (PEDOT:PSS) fibers were produced by using a simplified wet spinning process. Optimum wet spinning conditions were shown to produce fibers with good mechanical and electrical properties. The fiber spinning set-up involved the extrusion of the spinning solution from the top and passing into a vertical glass column containing the coagulation bath. Use of isopropanol as a coagulation bath reduced the coagulation rate during spinning.[18]

Continuous graphene fibers were produced by the wet spinning method followed by chemical reduction. Graphene is among the most interesting and attractive carbon-based materials. Graphene oxide (GO) dopes were injected into a coagulation bath of hexadecyltrimethyl ammonium bromide (CTAB). The CTAB solution was selected as the coagulation bath due to its positive charge. A 1.6 m long fiber could be drawn from the bath. This was followed by highly-efficient chemical reduction.[19]

In conclusion, wet spinning is a comparatively cost-effective production method. The drawbacks of low spinning speeds have recently been eliminated. Additionally, recent developments have shown that wet-spun fibers have wide application in the area of high-tech textiles.

9.7 References

1. V. B. Gupta and V. K. Kothari (1997), *Manufactured Fiber Technology*, Chapman and Hall, London, UK.
2. R. B. Chavan and A. K. Patra (2004), 'Development and processing of lyocell', *Indian Journal of Fiber and Textile Research*, **29**, December, 483–492.
3. C. Woodings (Ed.) (2001), *Regenerated Cellulose Fibers*, Woodhead Publishing Limited, Cambridge, England.
4. H. R. Mauersberger (1947), *Textile Fibers*, John Wiley and Sons, Inc., New York, USA.
5. T. Nakajima (Ed.) (2000), *Advanced Fiber Spinning Technology*, Woodhead Publishing Limited, Cambridge, England.
6. P. V. Vidyasagar (Ed.) (2000), *Encyclopedia of Textiles*, Mittal Publications, New Delhi, India.
7. H. Aizawa and F. Ishida (1962), Process and apparatus for stretch-spinning cuprammonium rayon, United States Patent 3049755.
8. J. Robertson (1999), *Forensic Examination of Fibers*, CRC Press, London.
9. *Lyocell Fibers: Production, Properties, End-Uses*, Melliand International, **1**, 6–8, (1995).
10. D. Loubinoux and S. Chounis (1985), An experimental approach of spinning new cellulose fibers with NMMO (N-Methylmorpholine-oxide) as solvent of cellulose, Lenzinger Berichte, August, 105–109.
11. A. Aghanouri, A. Zadhoush and M. Haghighat (2009), 'Effect of dry-wet spinning on the photooxidative degradation of acrylic fibers', *Journal of Applied Polymer Science*, **111**, 945–952.
12. S. B. Warner (1995), *Fiber Science*, Prentice Hall, New Jersey.
13. A. Ziabacki (1976), *Fundamentals of Fiber Formation*, John Wiley and Sons, New York.
14. A. Rende (1972), 'A new approach to coagulation phenomena in wet-spinning', *Journal of Applied Polymer Science*, **16**(3), 585–594.

15. B. Ji (2010), 'Study on the mass transfer process in pan wet-spinning', *Materials Science Forum*, **650**, 336–342.
16. C.-K. Liu, J. A. Cuculo and B. Smith (1989), Coagulation studies for cellulose in the ammonia/ammonium thiocyanate (NH_3/NH_4SCN) direct solvent system, *Journal of Polymer Science Part B: Polymer Physics*, **27**(12), 2493–2511.
17. R.N.R. Araiza, C. Rochas, L. David and A. Domard (2008), 'Interrupted wet-spinning process for chitosan hollow fiber elaboration', *Macromolecular Symposia*, **266**(1), 1–5.
18. R. Jalili, J. M. Razal, P. C. Innis and G. G. Wallace (2011), 'One-step wet-spinning process of poly(3,4-ethylenedioxythiophene): poly(styrenesulfonate) fibers and the origin of higher electrical conductivity', *Advanced Functional Materials*, **21**(17), 3363–3370.
19. H. P. Cong, X. C. Ren, P. Wang and S. H. Yu (2012), 'Wet-spinning assembly of continuous, neat and macroscopic grapheme fibers', *Scientific Reports*, **2**, 613.

10 Dry spinning of synthetic polymer fibers

Y. IMURA, New Jersey Institute of Technology, USA,
R. M. C. HOGAN, Omni Tech International, USA and
M. JAFFE, New Jersey Institute of Technology, USA

DOI: 10.1533/9780857099174.2.187

Abstract: Dry spinning is the fiber formation process that transforms a high vapor pressure polymer solution to a solid fiber by controlled fiber evaporation in the spinline. The key variables in dry spinning are heat transfer, mass transfer and stress on the filament. The spinning fiber reaches its final velocity/final denier when the stress applied to the spinning filament is no longer sufficient to draw the fiber to a smaller diameter. Solvent evaporation proceeds from the fiber surface inwards, often leading to a skin-core structure with trapped solvent in the fiber interior. Diffusion of this residual solvent can lead to the collapse of cylindrical fiber to the characteristic 'dogbone' shape often associated with dry spun fibers. This shape is also responsible for improved comfort and fabric cover. Most important of the dry spun fibers and cellulose acetate (CA), cellulose triacetate(CTA) and acrylics. Other fibers that are wet spun include spandex, polybenzimidazole (PBI) and polyvinylchloride (PVC).

Key words: solvent spinning, polymer solution, vapor pressure, skin-core, solvent diffusion through filament wall, dogbone, cellulose acetate, cellulose triacetate, acrylic, spandex, polybenzimidazole, polyvinylchloride.

10.1 Introduction

There are three major methods for producing polymeric fibers:

melt spinning
wet spinning
dry spinning.

The different processes for melt, dry, and wet spinning are shown in Fig. 10.1. Most commercial synthetic fibers are produced by the melt spinning process, which is a relatively simple and low-cost technique compared to wet or dry spinning. However, for polymers without a stable melt phase, spinning

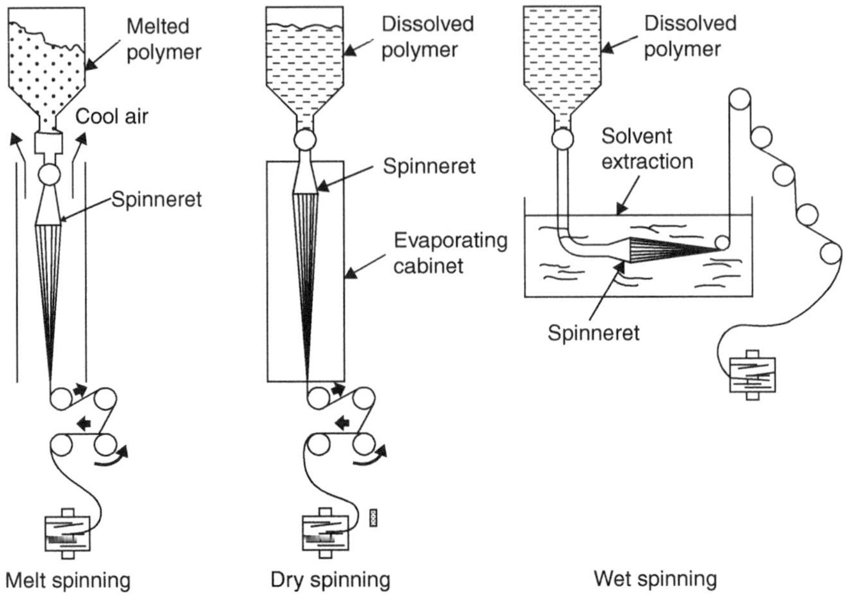

10.1 Melt, dry and wet spinning.

the fiber from a solution is required, a process known as solution spinning. There are two types of solution spinning:

- If a highly volatile solvent (high vapor pressure) can be employed, such that the fiber forms by solvent evaporation during spinning, the process is known as dry spinning.
- If the only potential polymer solvents are of low volatility (low vapor pressure), the fiber needs to be spun into a non-solvent bath to remove the spinning solvent and also to coagulate the fiber, in which case the process is known as wet spinning.

Dry spinning is used when the polymer is vulnerable to thermal degradation, cannot form thermally stable or viscous melts, or when specific surface characteristics of the filament are required (Gou and McHugh, 2004).

This chapter reviews key steps in the dry spinning process:

- Characteristics of dry spun fibers
- Key variables in dry spinning
- Dry spinning of particular fiber types: acrylics, cellulose acetate and triacetate, polybenzimidazole (PBI), spandex and polyvinyl chloride (PVC)
- More recent developments in dry spinning.

10.2 The dry spinning process

Dry spinning is used to produce fibers from polymers. such as cellulose acetate, cellulose triacetate, polyvinyl chloride, polymers and co-polymers of vinyl chloride, and acrylonitrile. Fibers that are dry spun include acrylics, spandex, PBI, and PVC. The basic dry spinning process begins with dissolving the polymer in an organic solvent. Polymers for dry spinning must be soluble in a volatile solvent such as ether or acetone. This solution is blended with additives and filtered, producing a viscous polymer solution. This solution is called 'dope.' The dope (which needs to be of much lower viscosity than for melt spinning) is filtered, de-aired, and pre-heated. It is then pumped through filters to achieve the right consistency.

The dope is then extruded in a spinning tube (also known as a spinning cell or evaporation chamber). In the extrusion process, fiber-forming substances in the solution are forced through the fine orifices in a metallic plate called a spinneret (or jet) at a controlled rate. The exiting jets of the polymer solution come into contact with a stream of hot gas. The gas may be air or an inert gas: air in the case of acetate, nitrogen (N_2) in the case of polyurethane. The solvent vaporizes in this gas stream, increasing the polymer concentration in the filament and thus solidifying it without the need for further drying. As the viscous filament streams enter the gas flow, the solvent first evaporates from the surface, forming a solid skin and then, with further evaporation during the downward passage through the gas flow, fully solidified filaments, resulting in the characteristic dumbbell/dog bone shape of dry spun fibers (discussed in more detail in Section 10.3).

The solidified filaments are drawn off by rotating rolls (or godets) and put onto bobbins or bales. Simultaneous stretching by down-drawing can also be applied in the same way as in melt spinning, enabling orientation of the polymer chains along the fiber axis. Filaments are typically combined into a yarn and wound onto a package by a ring and traveler system, which inserts a small degree of twist to hold the filaments together.

The solution concentration in the spinning dope ranges from 15% to 40% with corresponding viscosities ranging from 300 to 5000 P. The temperature of the inlet gas may range from 100°C to 250°C. Typical spinning velocities range from 10^2 to 10^3 m/min. Production rates typically range from 100–800 mpm. In a solvent spinning process such as dry spinning, the solvent is generally recovered. Capture and recovery of this solvent is important for both cost and environmental reasons. Recovery systems vary by polymer and solvent type, but include gas adsorption, gas absorption, condensation, and distillation methods.

The process is more complex than melt spinning, because production of the filament is a function both of temperature and dope concentration, which is coupled to the mass transfer mechanisms of solvent evaporation and diffusion. As a result, dry spun fibers are typically more expensive to produce than melt spun fibers.

10.3 Characteristics of dry spun fibers

Dry spun fibers have a characteristic dumbbell or dog bone shape compared to wet spun fibers, which are mainly round or kidney bean shaped. This is the result of the solvent diffusion and evaporation stages that lead to polymer solidification. As the solvent evaporates from the polymer solution, the outer part of the forming fiber solidifies before the inner part. This causes radial inhomogeneity within each filament. The outer part collapses inwards to produce the characteristic shape (see Fig. 10.2). Dry spinning also typically produces fibers that have rough rather than smooth surfaces compared to fibers produced by melt extrusion.

Because of the composition and characteristic cross-section of the fiber, which gives it a relatively large surface area, fabrics made from dry spun fibers typically have a high bulk-to-weight ratio and better cover. The characteristic cross-section means that more air is trapped between fibers, producing better insulating properties in the resulting fabric. In addition, the shape of dry spun fibers results in good dimensional stability with uniform breaking elongation, good tenacity and resiliency but a reduced modulus. These fiber properties are well-suited for interior textile applications such as carpets which require a strong, wear resistant fiber.

The other major difference between dry spun fibers and those produced by other techniques such as wet or melt spinning is the absence of the gel-state phase in dry spinning, which results in significant differences in physical properties. This can be seen, for example, in studies of the dry spinning of polyamide (PA) fibers reviewed by Vasanthan (2009). Tensile properties of dry spun PA fibers have been found to be superior to wet spun PA fibers (Hancock *et al.*, 1977). A separate study showed that dry spinning of PA6 fibers resulted in higher tensile strength compared to fibers prepared by melt spinning (Gogolewski and Pennings, 1985). These differences have been related to better molecular orientation in the fibers during fiber formation and subsequent drawing, the use of high molecular weight polymers, and the high concentration of polymer to solvent used in dry spinning (Vasanthan, 2009).

10.2 Formulation of dry spun fibers.

10.4 Variables in dry spinning

There are a number of variables that control fiber quality in dry spinning (Ziabicki, 1976; Van Kerevelen and Nijenhuis, 2009). These include:

Polymer solution concentration and rheology
Die swell and skin friction in filament formation
Filament solidification.

As in all fiber spinning processes, the key engineering variables are temperature, T, stress, σ, and, in the case of solution spinning, mass transfer, β. Key to solution properties is the polymer molecular weight, M, which is a strong factor in both solution viscosity and solidification rate.

The following sections review these variables.

10.4.1 Polymer solution concentration and rheology

The dry spinning dope is a polymer solution. The presence of a solvent component adds to the more complex nature of the viscosity of the solution in comparison to that for polymer melts. The inclusion of the solvent component makes the viscosity of the polymer solution a function of the nature and concentration of the solvent and the temperature of the solution. Concentrated polymer solutions exhibit a property known as critical molecular mass, where for a plot of log η_0 vs log M, the slope undergoes a dramatic change when M exceeds a critical value M_{cr}. The effect of the molecular mass and concentration on the viscosity was initially determined by the power-law equation (Krevelen and Nijenhuis, 2009):

$$\eta_0 = Kc^{\infty} M^{\beta} \qquad [10.1]$$

where K, ∞, and β are constants. The effect of temperature on the viscosity of the concentrated solutions is similar to that for dilute solutions, where it is described by the Andrade equation:

$$\eta_S = B \exp\left(\frac{E_\eta}{RT}\right) \qquad [10.2]$$

where B = constant and E_η is the activation energy. The viscosity of the polymer solution is also dependent on the shear rate. At low shear rates, polymer solutions exhibit Newtonian behavior. Interestingly, as the shear rate is increased and the viscosity at first decreases, the curve approaches another constant value, known as the second Newtonian region. At the shear rates found in dry spinning (up to 10–10s^{-1}) the fluid behaves in a non-Newtonian manner (Ziabicki, 1976).

10.4.2 Die swell and skin friction

The hydrodynamics of the dry spinning fluid is affected by the die swell phenomenon. Die swell can affect both fiber strength and spinning efficiency (Deopora *et al.*, 2013). Non-Newtonian shear flow within the spinneret channel with radius, R_0, causes the fluid jet exiting the channel to contract axially, followed by an increase in radius to a value of R_{max}. During this radius expansion, the velocity decreases to conserve mass. The degree of swelling, called the die swell ratio, is $\chi = R_{max}/R_0$ and depends on the conditions of extrusion, temperature, and the geometry of the extrusion die (Ziabicki, 1976). In dry spinning, the free jet undergoes elongation between the spinneret and the take-up device. The elongational viscosity is assumed to be three times the zero-shear viscosity and is valid for low deformation rates. This viscosity is also known as the Trouton viscosity.

The forces acting on the spin line are given in the general force balance (Ziabicki, 1976):

$$F_{ext} + F_{grav} = F_{surf} + F_{in} + F_{aero} + F_{rheo} \quad [10.3]$$

where the terms are for the applied external tension, gravity, surface tension, inertia, air drag, and the rheological effects, respectively. The most significant departure from melt spinning takes place in the determination of F_{aero}, which is affected by the skin friction coefficients. Due to the mass transfer via evaporation between the polymer fluid jet and its ambient environment, the boundary layer thickness changes. This in turn affects the skin friction, heat transfer, and the mass transfer coefficient parameters of the system.

10.4.3 Solidification of the filament

Solidification of the dry spinning filament is due to the evaporation of the solvent component from the spin line. There are three mechanisms in the solvent removal process:

- flash vaporization
- diffusion within the spin line
- convective mass transfer from the filament surface to the surrounding gas medium.

Flash vaporization of the solvent occurs at the spinneret exit as a result of the decompression of the hot polymer solutions. This mechanism is governed by the relationship between the pressure of the solvent, P_s, over the polymer solution with the volume fraction of the solvent, c_s, as given by the Flory-Huggins theory (Ziabicki, 1976):

$$\ln\left(\frac{P_s}{P_s^0}\right) = \ln c_s + (1-c_s)(1-\zeta) + \chi_{12}(1-c_s)^2 \qquad [10.4]$$

where P_s^0 is the partial pressure of pure solvent, ζ is the ratio of the molecular solvent volume to the molecular polymer volume, and χ_{12} is the Huggins polymer-solvent interaction parameter.

A particularly important mechanism of solvent transfer in dry spinning, and one that has been shown to be the rate controlling factor in filament solidification, is the process of diffusion. The diffusion process for a binary polymer-solvent system is described by the Fick equation:

$$\frac{\partial c_s}{\partial t} = r^{-1}\left(\frac{\partial}{\partial r}\right)\left(D^* r \frac{\partial c_s}{\partial r}\right) \qquad [10.5]$$

where D^* is the mutual diffusion coefficient and r is the radius.

Of equal importance in the mass transport process is the convection of the solvent into the gas phase. The boundary condition for diffusion in this convective case is given by

$$D^*\left(\frac{\partial c_s}{\partial r}\right)_{r=R} = -\beta\frac{(P_{s,R} - P_{s,\infty})}{(P - P_{s,\infty})} = \frac{-j_s}{\rho} \qquad [10.6]$$

where β is the convective mass transfer coefficient, $P_{s,R}$ and $P_{s,\infty}$ are the partial pressures of the solvent on the filament and gas sides, respectively, and P is the total pressure in the gas phase.

10.5 Dry spinning of acrylic fibers

The importance of some of these process issues relating to polymer solution rheology and solidification can be seen in the dry spinning of acrylics (Cox, 2005). An acrylic fiber is one in which the fiber-forming substance is any long chain synthetic polymer composed of at least 85% by weight of acrylonitrile (AN) units (-CH_2-CH(CN)-). Acrylic fibers are made from the polymer polyacrylonitrile (PAN), a synthetic, semicrystalline, organic polymer

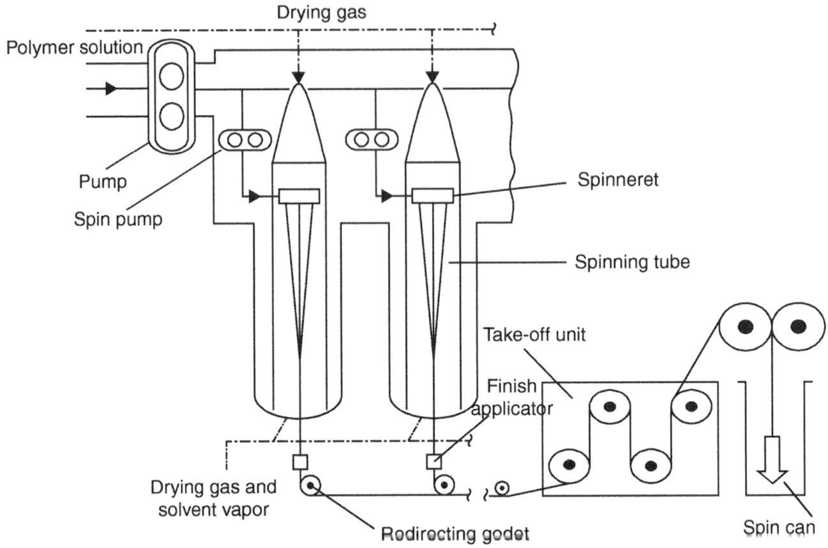

10.3 Dry spinning of acrylic fibers.

resin with a linear formula $(C_3H_3N)_n$. Typical comonomers are vinyl acetate $(C_4H_6O_2)$ and methyl acetate $(C_4H_6O_2)$. Acrylics are solvent-produced fibers using either wet or dry spinning processes.

A typical dry spinning unit for acrylic fibers is shown in Fig. 10.3. The polymer is dissolved in solvents such as N,N′-dimethylformamide (C_3H_7NO) (known as DMF) or aqueous sodium thiocyanate (NaSCN), at concentrations of 28–30% (Cox, 2005). DMF is the main solvent used in dry spinning because of its low boiling point and low cost (Masson 1995). The polymer is first dispersed in the cold solvent to form a slurry. It is then dissolved by application of heat and shear in a dope mixer or heater. A typical dope temperature is about 90°C. The dope is usually filtered and de-aerated prior to fiber spinning. It is passed through a series of filters of increasing fineness as it approaches the spinneret. Filtration pressure also removes air, resulting in an air-free dope at the end of filtration. Additives can be introduced during filtration, such as titanium dioxide, which acts as a delustrant, and carbon black, which acts as a pigment. As it passes through the final filters, the dope is heated up to 130–140°C. The pressure of the dope entering the spinning tube containing the spinneret is typically 0.8–1.5 MPa. Before reaching the spinnerets, pigments can be added. This leads to colored fibers with excellent fastness properties.

A typical spinneret is a ring made from stainless steel alloy and containing up to 2800 holes. Hole diameters are usually in the region of 0.1–0.3 mm. Bowl or hat-shaped spinnerets can also be used. Ring spinnerets have the advantage that the hot gases used to evaporate the solvent can be introduced through the center of the ring as well as from the outside, whereas it is possible to introduce gases only from the outside with other designs.

As the polymer leaves the spinneret it encounters hot gas at around 200–350°C, evaporating most of the solvent. Between 5% and 25% of the DMF is retained in the undrawn fiber to help orientate the molecular chains in the following drawing process (Kim, 2009). The quantity of solvent retained is dependent on temperature, column length, and take-up speed. A typical gas mixture is air mixed with nitrogen, which reduces the risk of explosions as well as potential yellowing of fibers. Gas flows can be co-current or counter-current. In co-current systems, the gas enters the spinning tube at the top close to the spinneret and is removed at the bottom through suction. In counter-current systems, the gas enters at the bottom of the tube and is removed at the top. Hot gas laden with solvent is removed so that the solvent can be recovered and re-used. In co-current systems, spinning tubes need to be operated under vacuum to prevent gas escaping with the fibers at the bottom.

A typical dry spinning line is divided into two units, each containing up to 35 spinning tubes. Output of fibers from each tube is 8 to 15 kg/h. The resulting fiber is taken up by the godet, and an oil finish may be applied before the fiber is collected in spin cans. The resulting fiber will then require further processing to improve its tensile and elastic properties. These processes include drawing, washing, crimping, and cutting.

Those involved in dry spinning of acrylics refer to a 'spinnability window,' i.e. the conditions required to produce a fiber of the right quality. As noted in the preceding section, polymer solution viscosity and temperature are key factors. A polymer solution viscosity that is too high or too low can lead to broken filaments, as can too high a polymer solution temperature. As also noted earlier, the rate of evaporation of the solvent is another key factor. If evaporation is too rapid, vapor bubbles can form in the filaments, reducing filament strength (Cox, 2005).

Because of the composition and characteristic cross-section of the fiber, fabrics made from acrylics have a high bulk-to-weight ratio. This gives them insulating properties, and makes them popular as substitutes for wool fibers. Because acrylic fibers are thermoplastic, fabrics may be heat-set for wrinkle resistance and to provide permanency to pleats. Acrylic fabrics have low moisture absorbency and dry relatively fast. In general, acrylic fibers are resistant to the degrading effects of ultraviolet rays in sunlight and to a wide range of chemicals and fumes. Acrylic fibers have relatively poor flame resistance compared with other fibers. Static and pilling can be a problem.

Acrylic fiber applications include floor coverings, blankets, throws, awnings, outdoor furniture, and apparel uses such as suits, pile fabrics, coats, sweaters, collars, linings, dresses, and shirts. 75% of acrylic fibers are used in apparel, 20% in home furnishing, and 5% in industrial end-uses. PAN fibers are the chemical precursors of high-quality carbon fibers. PAN is first thermally oxidized in air at 230°C and then carbonized above 1000°C in inert atmosphere to make carbon fibers. These fibers are found in high-tech applications for civil and military aircraft components, solid propellant rocket motors, pressure vessels, as well as sports equipment such as fishing rods, golf clubs, and tennis rackets.

Modacrylic (MAC) fibers are defined as those in which the fiber-forming substance is any long chain synthetic polymer composed of less than 85% but at least 35% by weight of acrylonitrile units. As with acrylics, both wet and dry spinning are used. They are flame retardant, do not combust, and have superior resistance to chemicals. They have relatively high durability, low tenacity, and moderate abrasion resistance. Fibers are resilient to mildew and sunlight, can be washed and dried, but should not be exposed to temperatures above 120°C or steamed because they are more heat sensitive. The principal end-uses of modacrylic fibers are in applications where environmental resistance or flame retardancy is needed, e.g. in carpets or protective gear. The low density of the fibers makes them suitable for lightweight protective clothing. They are also used in fleece, knit pile fabrics, paint rollers, stuffed toys, wigs, fake fur draperies, carpets, and filters.

10.6 Dry spinning of cellulose acetate and triacetate

In general, dry spinning of cellulose acetate is a cleaner and faster process than wet spinning and requires less labor (Hearle and Woodings, 2001). The cellulose for acetate production is reacted with acetic acid and acetic anhydride in the presence of sulfuric acid. It undergoes partial hydrolysis to remove the sulfate and a sufficient number of acetate groups to give the desired product properties. The anhydro glucose unit is the functional repeating cellulose structure and has three hydroxyl groups, which react to form acetate esters. Once the acetic anhydride is added, the acetylation reaction occurs. Temperature control of the process is critical to prevent cellulose degradation. The end result is a viscous liquid. Testing for intrinsic viscosity or degree of polymerization is done to determine when to stop the reaction. The next stage is hydrolysis. When the desired acetyl value is achieved, the hydrolysis is stopped by the addition of sodium, calcium, or magnesium salt solutions. Following acetylation, the resultant product is precipitated, purified, and dried to produce cellulose acetate flakes.

The flake is then dissolved in acetone and water to prepare the concentrated spinning solution or dope at a concentration of 15–30% solids. As with acrylic fibers, titanium oxide may be added as a delustrant, as well as pigments to color the fibers. The dope is subsequently filtered and de-aerated using varied filtration technologies to remove impurities.

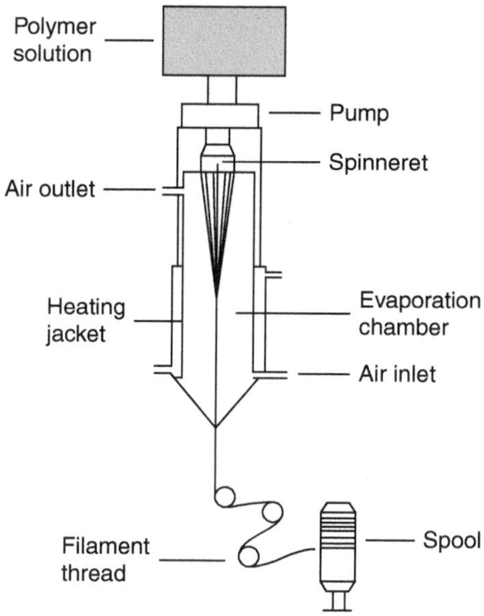

10.4 Dry spinning of cellulose acetate fibers.

After filtration, the highly viscous 'dope' solution is extruded through spinnerets into a column of warm air in which the acetone is evaporated, leaving solid continuous filaments of cellulose acetate (see Fig. 10.4). The dope typically descends vertically for 2–5 m against a counter-current of hot air. The temperature, moisture content and velocity of the air are important parameters. The spinneret holes are typically 50–100 microns (Hearle and Woodings, 2001). The number of holes, and the hole dimensions, vary based on product. Round holes in spinnerets produce an almost round crenulated filament (Fig. 10.5). Holes with different shapes can also be used. A stretch of around 100% is imposed on the yarn as it is drawn down from the spinneret. This stretch causes molecular orientation to develop in the plastic filaments in the upper part of the tube, providing the level of strength required for most textile applications. The evaporated acetone is recovered using a solvent recovery system and used to prepare additional spinning solution. For acetone the usual recovery process consists of adsorption by activated carbon and re-distillation.

A guide roll at the bottom of the spinning tube directs the fibers out to a collection device. For staple fiber, the filaments from numerous spinnerets are combined into tow form, crimped, cut to the required length, and packaged in bales. A take-up bobbin is used to insert a small amount of twist. Take-up speeds are typically between 200 and 400 m/min but can be as high as 1000 m/min. For filter tow the filaments are crimped, dried, and packaged in bales.

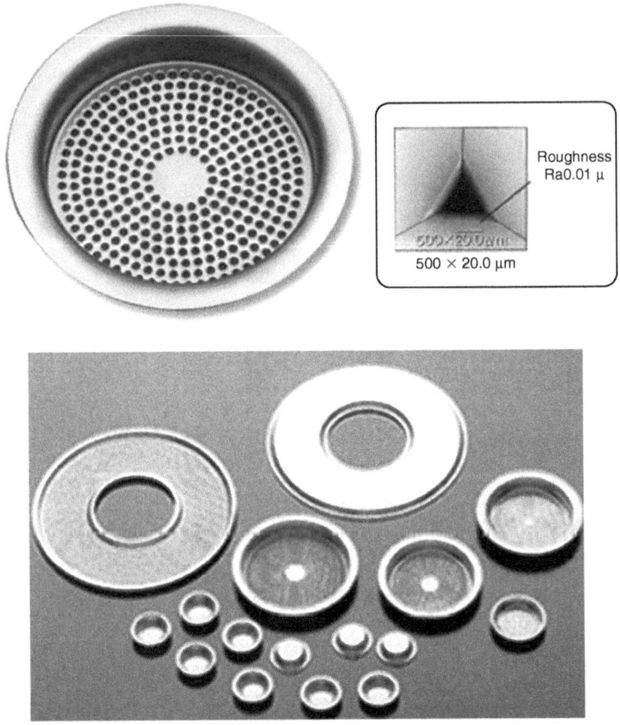

10.5 Spinnerets for dry spinning of cellulose acetate fibers.

The linear density (tex or denier) of the filaments depends on the rate of mass flow through the spinneret holes. Common filament sizes are around 3.7 denier (4.1 dtex). Yarns range from 55 to 600 denier (60.5–660 dtex). Fibers are relatively weak, extensible, and have low stiffness with poor elastic recovery above about 5% extension. Moisture absorption of fibers is high. Because of the characteristic fiber shape, which helps to trap air, cellulose acetate is also an excellent insulator. Typical industrial applications of acetate fibers include apparel, ribbons, cigarette filters, ink reservoirs, upholstery, drapes, and high absorbency product (diapers). Cigarette filter tow accounted for over 90% of the world consumption of acetate fibers in 2011. Fibers are hydrophilic, have breaking tenacities of 1.2–1.4 gpd, but rapidly loose strength when wet.

Triacetate uses dichloromethane and methanol as solvents. It is increasingly used as the film for LCD displays. It is very popular in films because of its excellent optical clarity, glossy surface characteristic, and resistance to grease. It is, however, prone to degradation. Fabrics of triacetate are shrink and wrinkle resistant. They are easily washable at high temperatures because they have higher heat resistance than acetate fabrics, and can be safely ironed at higher temperatures. Triacetate fabrics that have been properly heat-set (usually after dyeing) have improved ease-of-care characteristics because of a change in the crystalline structure of the fiber.

10.7 Dry spinning of other fibers: polybenzimidazole (PBI), spandex and polyvinyl chloride (PVC)

Dry spinning is also the method of choice for several other polymers including spandex, polyvinyl chloride (PVC) and polybenzimidazole (PBI). These polymers share the characteristic of thermally degrading prior to/during melting and solubility in high volatility solvents. In all cases it is the cost-effectiveness of dry spinning that leads to its adoption as the fiber formation method of choice.

10.7.1 Polybenzimidazole (PBI)

Polybenzimidazole or PBI is a manufactured fiber in which the fiber-forming substance is a long chain aromatic polymer having recurring imidazole groups as an integrated part of the polymer chain. PBI is a step growth polymerization process from 3,3′,4,4′-tetraaminobiphenyl and diphenyl isophthalate ($C_{20}H_{14}O_4$). A step growth polymerization refers to a type of polymerization mechanism in which bi-functional or multifunctional monomers form first dimes, then trimers, oligomers and eventually long chain polymers. The polymer is dissolved in solution with dimethylacetamide ($CH_3C(O)N(CH_3)_2$) and dope is then dry spun.

PBI is a high performance fiber used in firefighter, astronaut, racing car driver, and welder apparel. Fiber properties feature low static electricity buildup and low electrical conductivity. Its chemical resistance is good except against acids. The natural color of PBI is a gold-khaki shade, but it can be dyed to almost any medium to dark shade with conventional basic dyes after a caustic pretreatment. Fibers will not mildew and are abrasive resistant. A particularly important characteristic is heat resistance. Fibers do not melt in the presence of heat. They also do not ignite or smolder, retaining fiber integrity to 540°C, and only starting to degrade at 760°C. The fiber and fabrics from PBI retain their flexibility, dimensional stability, and significant strength without embrittlement, even when exposed to flame or extreme heat. PBI has high moisture regain and low modulus with comfort properties similar to cotton.

10.7.2 Spandex

Spandex is a synthetic fiber known for its elasticity. The term 'spandex' is the acronym in the USA, while the term 'elastane' is used in the European Union. It has been defined as a manufactured fiber in which the fiber-forming substance is a long chain synthetic polymer composed of at least 85% of segmented polyurethane. The fibers can be produced by several methods: melt, wet, dry, and reaction spinning. Dry spinning accounts for over 90% of production.

A prepolymer is required for dry spinning, as it is for other spinning techniques. Manufacture of spandex fibers begins with prepolymer production. The prepolymer is made up of long chain polyurethane, which is produced by reacting polyester with a diisocyanate monomer. This prepolymer is reacted with diamine. The spinning solution is pumped into a cylindrical spinning cell and extruded through the spinneret. The finer cells pass through the cell, which is heated in the presence of nitrogen and solvent gas. The liquid polymer reacts chemically and forms solid fibers. The polymer is extruded in a multiplicity of fine filaments to form a monofilament. The structure and properties of fibers is significantly influenced by mass flow rate through the spinnerets and take-up speed (Kotani *et al.*, 1992). Fiber bundles are further treated with finishing agents to prevent fibers from adhering to each other.

Spandex is lighter in weight, more durable and suppler than conventional elastic threads, and can be repeatedly stretched over 500% without breaking and still recover instantly to its original length. Fibers do not suffer deterioration from oxidation, and are not damaged by body oils, perspiration, lotions, or detergents. Stabilizers are added to protect fibers from heat, atmospheric contaminants, and chlorine. To protect fibers against light degradation, UV (ultraviolet) screeners such as hydroxybenzotriazoles are added. Spandex is white in color when first produced. Colorants are added later to improve its visual appearance. Normally dispersed and acid dyes are used. If spandex fibers are interwoven with other fibers like nylon or polyester then special dyeing methods are used. Major spandex uses include athletic wear, foundation garments, gloves, hosiery, orthopedic braces, hose, webbing, and stretchable clothing. It is blended quite often with polyester, to provide garments with a good look and feel but with added comfort.

10.7.3 Polyvinyl chloride (PVC)/vinyon

PVC or vinyon is defined as a manufactured fiber in which the fiber-forming substance is any long chain synthetic polymer composed of at least 85% by weight of vinyl chloride units ($-CH_2-CHCl$). Vinyl chloride is a chemical obtained from ethylene and chlorine. Most PVC fibers are produced by dry spinning, though some are produced by melt spinning. Dry spinning solvents used are acetone-benzene solution or acetone-dioxide solutions. PVC fibers have excellent moisture permeability. Moisture absorption is less than 0.5% and moisture regain is less than 0.1%. PVC is an outstanding flame retardant fiber because of its chlorine containing chemical structure. PVC does not ignite but softens at low temperatures (55°C). PVC also has excellent heat retention and negative electrostatic properties.

10.8 Conclusion and future trends

This chapter has reviewed the key processes in dry spinning, characteristics of dry spun fibers and applications in a range of synthetic fibers. The versatility of the process, suggested by the different fibers that have been successfully produced by dry spinning, suggests its potential for new applications. More recently, dry spinning has been used to spin carbon nanotube (CNT) fibers (Ghemes *et al.*, 2012). Carbon nanotubes have been one of the most widely investigated types of nanomaterial due to their outstanding electrical, thermal and mechanical properties. A range of spinning techniques have been used, including solution spinning (Zhang *et al.*, 2004; Shaffer and Sandler, 2007). Dry spinning techniques for CNTs were first developed by Windle, who referred to the process as 'direct spinning' (Li *et al.*, 2004). The technique involved injecting a liquid carbon mixture, plus catalyst and carrier gas, into a high-temperature electric furnace to form fibers that were then drawn from the bottom. Ghemes *et al.* (2012) describe growing carbon nanotubes using chemical vapor deposition (CVD) that were dry spun from vertically aligned arrays of carbon nanotubes to produce fibers with excellent mechanical properties, including high tensile strength (up to 500 MPa). Other researchers have also found that dry spinning improves the tensile strength of CNT fibers and yarn (Tran *et al.*, 2010). Dry spinning of CNT fibers has achieved a spinning rate of 20–30 m/min., suggesting it is suitable for mass production of CNT fibers (Choo *et al.*, 2012). Developments such as these indicate the continuing value and evolution of this method of spinning.

10.9 References

Choo H., Jung, Y, Jeong Y. and Kim, H. (2012), 'Fabrication and applications of carbon nanotube fibers', *Carbon Letters*, **13**(4), 191–204.

Cox, R. (2005), 'Acrylic fibres', in McIntyre, J. (ed.), *Synthetic Fibres*, Woodhead Publishing Limited, Cambridge.

Deopora, B., Chatterjee, A. and Padaki, N. (2013), 'Process control in the manufacturing of synthetic textile fibres', in Majumdar, A., Das, A., Alagirusamy, R. and Kothari, V. (eds.), *Process Control in Textile Manufacturing*, Woodhead Publishing Limited, Cambridge.

Ghemes, A., Muramutsu, J., Minami, Y. and Okada, M. (2012), 'High performance carbon nanotube fibers spun from long multi-walled carbon nanotubes', *Journal of Advanced Research in Physics*, **3**(1), 1–3

Gogolewski, S. and Pennings, A. (1985), 'High-modulus fibers of nylon-6 prepared by a dry-spinning method', *Polymer*, **26**, 1394–1400.

Gou, Z. and McHugh, A. J. (2004), 'Dry spinning of polymer fibers in ternary systems', *International Polymer Processing*, **19**(3), 244–253.

Hancock, T., Spruiell, J. and White J. (1997), 'Wet spinning of aliphatic and aromatic polyamides', *Journal of Applied Polymer Science*, **21**, 1227–1247.

Hearle, J. and Woodings, C. (2001), 'Fibres related to cellulose', in Woodings, C. (ed.), *Regenerated Cellulose Fibres*, Woodhead Publishing Limited, Cambridge.

Kim, H. (2009), 'The structure of acrylic, polyvinylalcohol and polyvinylchloride fibers', in S. Eichhorn, J. Hearle, M. Jaffe and T. Kikutani (eds.), *Handbook of Textile Fibre Structure Volume 1: Fundamentals and Manufactured Polymer Fibres*, Woodhead Publishing Limited, Cambridge.

Kotani, T, Hayashi, S., Ishihara, H. and Matsui, T. (1992), 'Structure and properties of dry-spun segmented polyurethaneurea elastic fibers', *Journal of Macromolecular Science, Part B: Physics*, **31**(1), 65–86.

Li, Y., Kinloch, I. and Windle, A (2004), 'Direct spinning of carbon nanotube fibers from chemical vapor deposition synthesis, *Science*, **304**, 276—278, DOI: 10.1126/science.1094982.

Masson, J. (1995), *Acrylic Fiber Technology and Applications*, Marcel Dekker, New York.

Shaffer, M. and Sandler, J. (2007), 'Carbon nanotube and nanofibre reinforced poly mer fibres', in P. Brown and K. Stevens (eds.), *Nanofibers and Nanotechnology in Textiles*, Woodhead Publishing Limited, Cambridge.

Tran, C., Humphries, W., Smith, S. and Huynh, C. (2010), 'Improving the tensile strength of carbon nanotube spun yarns using a modified spinning process', Paper delivered at the 2010 International Conference on Nanoscience and Nanotechnology (ICONN), Sydney, Australia.

Van Krevelen, D.W and Nijenhuis, K. (2009), *Properties of Polymers Fourth Edition*, Elsevier B.V., Amsterdam.

Vasanthan, N. (2009), 'Polyamide fiber formation', in S. Eichhorn, J. Hearle, M. Jaffe and T. Kikutani (eds.), *Handbook of Textile Fibre Structure Volume 1: Fundamentals and Manufactured Polymer Fibres*, Woodhead Publishing Limited, Cambridge.

Ziabicki, A. (1976), *Fundamentals of Fibre Formation: the Science of Fibre Spinning and Drawing*, John Wiley and Sons, New York.

Zhang, M., Atkinson, K. and Baughman, R. (2004), 'Multi-functional carbon nanotube yarns by downsizing an ancient technology', *Science*, **306**, 1358–1361.

Index

FSC
www.fsc.org
MIX
Papier | Fördert
gute Waldnutzung
FSC® C083411

Kolibri 360 GmbH
Ferdinand-Jühlke-Straße 7
99095 Erfurt, Deutschland
produktsicherheit@kolibri360.de